The 5th Conference on Sustainability in Civil Engineering (CSCE)

The 5th Conference on Sustainability in Civil Engineering (CSCE)

Editors

Majid Ali
Muhammad Ashraf Javid
Shaheed Ullah
Iqbal Ahmad

Basel • Beijing • Wuhan • Barcelona • Belgrade • Novi Sad • Cluj • Manchester

Editors

Majid Ali
Department of
Civil Engineering
Capital University of Science
and Technology
Islamabad, Pakistan

Muhammad Ashraf Javid
Department of
Civil Engineering
Capital University of Science
and Technology
Islamabad, Pakistan

Shaheed Ullah
Department of
Civil Engineering
Capital University of Science
and Technology
Islamabad, Pakistan

Iqbal Ahmad
Department of
Civil Engineering
Capital University of Science
and Technology
Islamabad, Pakistan

Editorial Office
MDPI
St. Alban-Anlage 66
4052 Basel, Switzerland

This is a reprint of articles from the Proceedings published online in the open access journal *Engineering Proceedings* (ISSN 2673-4591) (available at: https://www.mdpi.com/2673-4591/44/1).

For citation purposes, cite each article independently as indicated on the article page online and as indicated below:

Lastname, A.A.; Lastname, B.B. Article Title. *Journal Name* **Year**, *Volume Number*, Page Range.

ISBN 978-3-0365-9202-2 (Hbk)
ISBN 978-3-0365-9203-9 (PDF)
doi.org/10.3390/books978-3-0365-9203-9

Cover image courtesy of Engr. Iqbal Ahmad

© 2023 by the authors. Articles in this book are Open Access and distributed under the Creative Commons Attribution (CC BY) license. The book as a whole is distributed by MDPI under the terms and conditions of the Creative Commons Attribution-NonCommercial-NoDerivs (CC BY-NC-ND) license.

Contents

About the Editors vii

Majid Ali, Muhammad Ashraf Javid, Shaheed Ullah and Iqbal Ahmad
Statement of Peer Review [†]
Reprinted from: *Eng. Proc.* **2023**, *44*, 8, doi:10.3390/engproc2023044008 1

Majid Ali, Muhammad Ashraf Javid, Shaheed Ullah and Iqbal Ahmad
Preface of the 5th Conference on Sustainability in Civil Engineering [†]
Reprinted from: *Eng. Proc.* **2023**, *44*, 9, doi:10.3390/engproc2023044009 3

Fakhar Hassan Shah, Omer Shujat Bhatti and Shehryar Ahmed
A Review of the Effects of Project Management Practices on Cost Overrun in Construction Projects [†]
Reprinted from: *Eng. Proc.* **2023**, *44*, 1, doi:10.3390/engproc2023044001 9

Fakhar Hassan Shah, Omer Shujat Bhatti and Shehryar Ahmed
Project Management Practices in Construction Projects and Their Roles in Achieving Sustainability—A Comprehensive Review [†]
Reprinted from: *Eng. Proc.* **2023**, *44*, 2, doi:10.3390/engproc2023044002 15

Qazi Umar Farooq and Muhammad Tayyab Naqash
Effectiveness of Mono Sand Piles in Soft Cohesive Ground [†]
Reprinted from: *Eng. Proc.* **2023**, *44*, 3, doi:10.3390/engproc2023044003 21

Waleed Nasir Khan, Syed Ghayyoor Hussain Kazmi and Anwar Khitab
Effect of Bio-Char of Santa Maria Feverfew Plant on Physical Properties of Fresh Mortar [†]
Reprinted from: *Eng. Proc.* **2023**, *44*, 4, doi:10.3390/engproc2023044004 25

Hamza Aamir, Kinza Aamir and Muhammad Faisal Javed
Linear and Non-Linear Regression Analysis on the Prediction of Compressive Strength of Sodium Hydroxide Pre-Treated Crumb Rubber Concrete [†]
Reprinted from: *Eng. Proc.* **2023**, *44*, 5, doi:10.3390/engproc2023044005 31

Maliha Mehar Qambrani, Fizza Mirza and Muhammad Habib
Comparative Seismic Response Analysis of a Multi-Storey Building with and without Base Isolators under High Magnitude Earthquake [†]
Reprinted from: *Eng. Proc.* **2023**, *44*, 6, doi:10.3390/engproc2023044006 37

Ahmed Kamal Subhani, Mohib Nisar and Anwar Khitab
Improvement of Early-Age Mechanical Properties of Cement Mortar by Adding Biochar of the Santa Maria Feverfew Plant [†]
Reprinted from: *Eng. Proc.* **2023**, *44*, 7, doi:10.3390/engproc2023044007 41

Fahad Ali, Hammad Azam and Muhammad Habib
Evaluation of Seismic Response of 3D Building Frame with and without Base Isolation Using Finite Element Analysis [†]
Reprinted from: *Eng. Proc.* **2023**, *44*, 10, doi:10.3390/engproc2023044010 47

Rehan Jamil, Hamidi Abdul Aziz and Mohamad Fared Murshed
Validation of Chlorine Decay Equation for Water Quality Analysis in Distribution Networks [†]
Reprinted from: *Eng. Proc.* **2023**, *44*, 11, doi:10.3390/engproc2023044011 53

Manail Shafqat, Muhammad Basit Khan and Hamad Hassan Awan
The Behavior of Pre-Treated Crumb Rubber and Polypropylene-Fiber-Incorporated Mortar Subjected to Elevated Temperatures [†]
Reprinted from: *Eng. Proc.* **2023**, *44*, 12, doi:10.3390/engproc2023044012 59

Asad Shafique, Ahsin Ihsan and Muhammad Faisal Javed
Efficiency and Sustainability: Enhancing Mortar Mixtures with Wastepaper Sludge Ash [†]
Reprinted from: *Eng. Proc.* **2023**, *44*, 13, doi:10.3390/engproc2023044013 65

Tariq Saeed and Muhammad Usman Arshid
Soil Improvement Using Waste Polyethylene Terephthalate (PET) [†]
Reprinted from: *Eng. Proc.* **2023**, *44*, 14, doi:10.3390/engproc2023044014 71

Nida Azhar, Farrukh Arif and Abdul Basit Khan
Framework for Energy Performance Measurement of Residential Buildings Considering Occupants' Energy Use Behavior [†]
Reprinted from: *Eng. Proc.* **2023**, *44*, 15, doi:10.3390/engproc2023044015 77

Shahzaib Farooq, Faheem Butt and Rana Muhammad Waqas
The Behavior of Retrofitted GPC Columns under Eccentric Loading [†]
Reprinted from: *Eng. Proc.* **2023**, *44*, 16, doi:10.3390/engproc2023044016 83

Samiullah Khan, Safeer Khattak and Hamza Khan
Composite Fibers in Concrete: Properties, Challenges, and Future Directions [†]
Reprinted from: *Eng. Proc.* **2023**, *44*, 17, doi:10.3390/engproc2023044017 87

Raheel Arif, Ammar Iqtidar and Safeer Ullah Khattak
Utilizing Corn Cob Ash and Bauxite as One-Part Geopolymer: A Sustainable Approach for Construction Materials [†]
Reprinted from: *Eng. Proc.* **2023**, *44*, 18, doi:10.3390/engproc2023044018 93

Muhammad Iqbal Bashir and Ayub Elahi
Micro Structural Study of Concrete with Indigenous Volcanic Ash [†]
Reprinted from: *Eng. Proc.* **2023**, *44*, 19, doi:10.3390/engproc2023044019 97

Javaria Mehwish, Katherine A. Cashell and Rabee Shamass
Flexure Response of Stainless-Steel-Reinforced Concrete (SSRC) Beams Subjected to Fire [†]
Reprinted from: *Eng. Proc.* **2023**, *44*, 20, doi:10.3390/engproc2023044020 103

About the Editors

Majid Ali

Dr. Majid is currently working as a Professor at the Capital University of Science and Technology, Islamabad, Pakistan. He has over 19 years of teaching, research, and professional industry experience. He obtained his B.Sc. in Civil Engineering with a gold medal, his M.Sc. with the highest position from the University of Engineering and Technology, Taxila, and his Ph.D. with five impact factor journal publications from the University of Auckland, New Zealand. He has published a total of 176 publications, including 36 Impact Factor journal papers with a cumulative ISI impact factor of more than 220. Before joining academia, for ten years he worked as a Structural Design Engineer with NESPAK (one of the leading consultants in Pakistan). His research interests include construction materials, fiber composites, and earthquake-resistant design.

Muhammad Ashraf Javid

Dr. Javid works as an Associate Professor at the Capital University of Science and Technology, slamabad, Pakistan. He completed his B.Sc. in Civil Engineering and M.Sc. in Transportation Engineering from the University of Engineering and Technology, Lahore, Pakistan, and earned his Ph.D. in Civil Engineering from Yokohama National University, Japan. He has 16 years of teaching and research experience in the civil engineering field and has worked at both national and international universities. His research areas mainly include passenger travel behavior, transportation demand management policies, the interaction between public transport services and customers, shared mobility and economy, and traffic safety. He has published a total of 55 journal publications and 21 conference proceedings, and his cumulative publication ISI impact factor is more than 70.

Shaheed Ullah

Engr. Shaheed Ullah works as lecturer at the Capital University of Science and Technology, Islamabad, Pakistan. He completed his B.Sc. in Civil Engineering and M.Sc. in Structural Engineering at the UET Peshawar and MCE NUST Campus, Risalpur, respectively. He has five years of teaching experience and has research expertise in construction materials.

Iqbal Ahmad

Engr. Iqbal works as a lecturer at the Capital University of Science and Technology, Islamabad, Pakistan. He is a Structural Engineer with a Master's degree in Structural Engineering. He has more than 9 years of teaching and industry experience. His research themes include finite element modeling, engineering materials, and numerical analysis.

Editorial

Statement of Peer Review †

Majid Ali *, Muhammad Ashraf Javid, Shaheed Ullah and Iqbal Ahmad

Department of Civil Engineering, Capital University of Science and Technology (CUST), Islamabad 44000, Pakistan; ashraf.javid@cust.edu.pk (M.A.J.); shaheed.ullah@cust.edu.pk (S.U.); iqbalahmad@cust.edu.pk (I.A.)
* Correspondence: majid.ali@cust.edu.pk; Tel.: +92-51-11155-5666 (ext. 354)
† All papers published in the volume are presented at the 5th Conference on Sustainability in Civil Engineering, Online, 3 August 2023.

In submitting the proceedings of the 5th Conference on Sustainability in Civil Engineering (CSCE) to *Engineering Proceedings*, the volume editors of the proceedings certify to the publisher that all papers published in this volume have been subjected to peer review administered by the volume editors. Reviews were conducted by expert referees to the professional and scientific standards expected of a proceedings journal.

- Type of peer review: double-blind
- Conference submission management system: through email
- Number of submissions received: 133
- Submissions sent for review: 72
- Number of submissions accepted: 56
- Acceptance rate (number of submissions accepted/number of submissions received): 42%
- The average number of reviews per paper: 3
- Total number of reviewers involved: 57
- Number of accepted submissions for MDPI: 18
- Any additional information on the review process: The papers which were received underwent a strict scrutiny process considering plagiarism (overall and single-source similarities should be present at a maximum of 19% and 4%, respectively), major formatting, completeness, referencing style, etc. Each paper needed at least two reviews. All modified papers again underwent a strict scrutiny process, positive consent was taken from technical committee members for papers presenting major concerns to ensure their quality before being accepted.

Citation: Ali, M.; Javid, M.A.; Ullah, S.; Ahmad, I. Statement of Peer Review. *Eng. Proc.* 2023, 44, 8. https://doi.org/10.3390/engproc2023044008

Published: 23 August 2023

Copyright: © 2023 by the authors. Licensee MDPI, Basel, Switzerland. This article is an open access article distributed under the terms and conditions of the Creative Commons Attribution (CC BY) license (https://creativecommons.org/licenses/by/4.0/).

Conflicts of Interest: The authors declare no conflict of interest.

Disclaimer/Publisher's Note: The statements, opinions and data contained in all publications are solely those of the individual author(s) and contributor(s) and not of MDPI and/or the editor(s). MDPI and/or the editor(s) disclaim responsibility for any injury to people or property resulting from any ideas, methods, instructions or products referred to in the content.

Editorial

Preface of the 5th Conference on Sustainability in Civil Engineering [†]

Majid Ali *, Muhammad Ashraf Javid, Shaheed Ullah and Iqbal Ahmad

Department of Civil Engineering, Capital University of Science and Technology, Islamabad, 44000, Pakistan; ashraf.javid@cust.edu.pk (M.A.J.); shaheed.ullah@cust.edu.pk (S.U.); iqbalahmad@cust.edu.pk (I.A.)
* Correspondence: majid.ali@cust.edu.pk
† All papers published in the volume are presented at the 5th Conference on Sustainability in Civil Engineering, Online, 3 August 2023.

1. Conference Overview

The Conference on Sustainability in Civil Engineering (CSCE), 2023, was organized by the Department of Civil Engineering of the Capital University of Science and Technology (CUST), Islamabad, Pakistan. The conference is held annually with the main focus on highlighting the sustainability-related aspects of research in civil engineering. It aims to provide a platform for civil engineers, from academia as well as industry, to share their practical experiences and different research findings in their relevant specializations. The major topics include green construction materials and structures, construction management for sustainable development, and resilient infrastructure and environment. This conference provides a remarkable opportunity for the academic and industrial communities to address new challenges, share solutions, and discuss future research directions in the field of civil engineering. Participants have the opportunity to attend the keynote lectures related to various recent trends in research networks in civil engineering. The conference accommodates several parallel sessions related to different specialties (Figure 1), where the researchers and engineers interact and enhance their understanding of sustainability in the context of civil engineering.

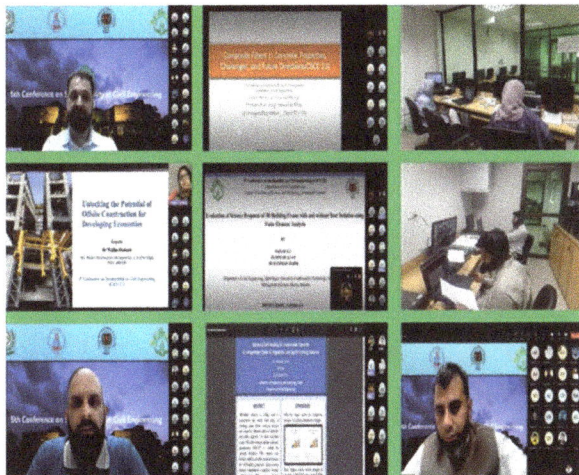

Figure 1. Different online sessions in the fifth edition of CSCE.

The fifth conference on sustainability in civil engineering was held on 3 August 2023. This year, we had nine wonderful and celebrated keynote speakers for this edition of the CSCE. We received 133 manuscripts from different countries around the world including UK, Ireland, Canada, Australia, Italy, Cyprus, China, Kazakhstan, Nigeria, Malaysia, KSA, and Pakistan. All papers underwent a comprehensive and critical double-blind review process. The review committee comprised 57 PhDs serving in industry and academia from the UK, Ireland, the USA, Australia, New Zealand, Singapore, Hong Kong, Poland, Italy, Chile, Malaysia, China, Oman, Bahrain, the KSA, and Pakistan. After the screening and review process, 56 papers were chosen for presentation at the conference.

We are grateful to all the reviewers and keynote speakers who have dedicated their precious time to share their expertise and experience. With this opportunity, we would also like to express our gratitude to everyone, especially all the faculty and staff at the Capital University of Science and Technology, for their great support and participation. In this regard, the participation and cooperation of all authors, presenters, and participants are also acknowledged, without whom this conference would not have been possible. Last but not least, we would like to express our appreciation to our advising and organizing committees, whose hard work and dedication have also made this day possible.

Majid Ali, Muhammad Ashraf Javid, Shaheed Ullah, and Iqbal Ahmad

2. About the Editors

- Majid Ali

Dr. Majid is currently working as a Professor at the Capital University of Science and Technology, Islamabad, Pakistan. He has over 19 years of rich teaching, research, and professional industry experience. He obtained his B.Sc. in Civil Engineering with a gold medal, his M.Sc. with the highest position from the University of Engineering and Technology, Taxila, and his Ph.D. with five impact factor journal publications from the University of Auckland, New Zealand. He has published a total of 176 publications, including 36 Impact Factor journal papers having a cumulative ISI impact factor of more than 220. Before joining academia, for ten years he worked as a Structural Design Engineer with NESPAK (one of the leading consultants in Pakistan). His research interests include construction materials, fiber composites, and earthquake-resistant design.

- Muhammad Ashraf Javid

Dr. Javid works as an Associate Professor at the Capital University of Science and Technology, Islamabad, Pakistan. He completed his B.Sc. in Civil Engineering and M.Sc. in Transportation Engineering from the University of Engineering and Technology, Lahore, Pakistan, and earned his Ph.D. in Civil Engineering from Yokohama National University, Japan. He has 16 years of teaching and research experience in the civil engineering field and has worked at both national and international universities. His research areas mainly include passenger travel behavior, transportation demand management policies, the interaction between public transport services and customers, shared mobility and economy, and traffic safety. He has published a total of 55 journal publications and 21 conference proceedings, and his cumulative publication ISI impact factor is more than 70.

- Iqbal Ahmad

Engr. Iqbal works as a lecturer at the Capital University of Science and Technology, Islamabad, Pakistan. He is a Structural Engineer with a Master's degree in Structural Engineering. He has more than 9 years of teaching and industry experience. His research themes include finite element modeling, engineering materials, and numerical analysis.

- Shaheed Ullah

Engr. Shaheed Ullah works as lecturer at the Capital University of Science and Technology, Islamabad, Pakistan. He completed his B.Sc. in Civil Engineering and M.Sc. in Structural Engineering at the UET Peshawar and MCE NUST Campus, Risalpur, respectively. He has five years of teaching experience and has research expertise in construction materials.

3. Technical Committee Members and Reviewers

Dr. Shunde Qin, Jacobs Sutton Coldfield, UK
Dr. Furqan Qamar, WSP, UK
Dr. Afaq Ahmed, University of Memphis, USA
Dr. Mohsin Shehzad, WSP, New Zealand
Dr. Claudio Oyarzo-Vera, UCSC College of Engineering, Chile
Dr. Munir Ahmed, DAR, KSA
Dr. Rabee Shamass, London South Bank University, UK
Dr. Mehran Khan, University College Dublin, Ireland
Dr. Piotr Smarzewski, Military University of Technology, Poland
Dr. Jincheng Liu, Nanyang Technological University, Singapore
Dr. Nicola Fontana, Universita Degli Studi del Sannio, Italy
Dr. Muhammad Shakeel, Hong Kong UST, Hong Kong
Dr. Anas Bin Ibrahim, MARA University of Technology, Malaysia
Dr. Noor Aina Misnon, Universiti Pertahanan Nasional, Malaysia
Dr. Li Li, Northwest A&F University, China
Dr. Mohsin Usman Qureshi, Sohar University, Oman
Dr. Umar Farooq, Islamic University of Madinah, KSA
Dr. Uneb Gazder, University of Bahrain, Bahrain
Dr. Irfan Yousuf, CC & EC, World Bank Group, Pakistan
Dr. Hassan Farooq, DD Consultancy Directorate, Pakistan
Dr. M Zia ur Rehman Hashmi, Global Change Impact Studies Centre, Pakistan
Dr. Hassan Nasir, WSSP, Peshawar, Pakistan
Dr. Shahid Nasir, FE Pvt. Ltd., Islamabad, Pakistan
Dr. Rao Arsalan Khushnood, Tunnelling Inst. of Pak. (TIP), Islamabad, Pakistan
Dr. Hammad Anis Khan, NUST, Islamabad, Pakistan
Dr. Imran Hashmi, NUST, Islamabad, Pakistan
Dr. Habib Ur Rehman, UET, Lahore, Pakistan
Dr. Noor Muhammad Khan, UET, Lahore, Pakistan
Dr. Ammad Hassan, UET, Lahore, Pakistan
Dr. Khurram Rashid, UET, Lahore, Pakistan
Dr. Qaisar Ali, UET, Peshawar, Pakistan
Dr. Amjad Naseer, UET, Peshawar, Pakistan
Dr. Ayub Elahi, UET, Taxila, Pakistan
Dr. Faisal Shabbir, UET, Taxila, Pakistan
Dr. Naveed Ahmad, (Transportation) UET, Taxila, Pakistan
Dr. Naveed Ahmad, (Geotech) UET, Taxila, Pakistan
Dr. Usman Ali Naeem, UET, Taxila, Pakistan
Dr. Jawad Hussain, UET, Taxila, Pakistan
Dr. Faheem Butt, UET, Taxila, Pakistan
Dr. Irshad Qureshi, UET, Taxila, Pakistan
Dr. Muhammad Usman Arshid, UET, Taxila, Pakistan
Dr. Farrukh Arif, NED, Karachi, Pakistan
Dr. Tahir Mehmood, COMSATS, Wah, Pakistan
Dr. Ahsen Maqsoom, COMSATS, Wah, Pakistan
Dr. Hassan Ashraf, COMSATS Wah, Pakistan
Dr. Adnan Nawaz, COMSATS, Wah, Pakistan
Dr. Muhammad Faisal Javed, COMSATS, Abbottabad, Pakistan
Dr. Rashid Farooq, IIU, Islamabad, Pakistan
Dr. Zeeshan Alam, IIU, Islamabad, Pakistan
Dr. Mudasser Muneer Khan, BZU, Multan, Pakistan
Dr. M. Shoaib, BZU, Multan, Pakistan
Dr. Anwar Khitab, MUST, Pakistan, Pakistan
Dr. Salman Ali, Suhail University of Lahore, Pakistan

Dr. Shoukat Ali Khan, Abasyn, Peshawar, Pakistan
Dr. Sabahat Hassan, HITEC University, Taxila, Pakistan
Dr. Jehangir Durrani, Iqra National University, Peshawar, Pakistan
Dr. Khursheed Ahmed, KIU, Gilgit, Pakistan

4. About the Keynote Speakers

Dr. Wajiha Shahzad *Massey University, New Zealand*	*Unlocking the Potential of Offsite Construction for Developing Economies*
Dr. Afaq Ahmed *University of Memphis, USA*	*State-of-the-Art Techniques for Structural Health Monitoring of Structures*
Dr. Hamza Farooq Gabriel *NUST, Islamabad, Pakistan*	*Sustainable Green Eco-Technologies for Wastewater Treatment*
Dr. Hossein Derakhshan *QUT, Australia*	*Seismic Vulnerability Assessment of Unreinforced Masonry Buildings*
Dr. Anwar Khitab *MUST, Mirpur, Pakistan*	*Carbon Storage in Building Materials for a Greener Future*
Dr. Imran Muhammad *Massey University, New Zealand*	*Prosperous Cities Transition—How Do Pakistani Cities Thrive in the Future by Adopting a Transit-Oriented Development (TOD) Model?*
Dr. Mizan Ahmed *Curtin University, Australia*	*Sustainable Use of High-Performance-Fiber-Reinforced Cementitious Composites*
Dr. Robert Evans *Nottingham Trent University, UK*	*The Use of Ground-Penetrating Radar (GPR) to Improve Sustainability in Pavement Engineering Projects*
Dr. Sajjala Sreedhar Reddy *University of Nizwa, Oman*	*Role of Civil Engineering in Sustainable Development*

5. About the Committees

Organizing Committee	Advisory Committee
Dr. Majid Ali Chair	**Dr. M. Mansoor Ahmed** *(Vice-Chancellor CUST Islamabad)* Patron
Dr. M. Ashraf Javid Co-Chair	**Dr. Imtiaz Ahmed Taj** *(Dean Faculty of Engineering)* General Advisor
Engr. Shaheed Ullah Conference Secretary-I	
Engr. Iqbal Ahmad Conference Secretary-II	**Dr. Ishtiaq Hassan** *(Head of Civil Engineering Department)* Principle Advisor

6. CSCE Industrial Partners for Its Fifth Edition

Global Climate Change Impact Studies Centre

Engineering & Design Consultants

City Engineering Consultants

MKAI Climate Consulting

ZAC Engineers

MARK Associates

Conflicts of Interest: The authors declare no conflict of interest.

Disclaimer/Publisher's Note: The statements, opinions and data contained in all publications are solely those of the individual author(s) and contributor(s) and not of MDPI and/or the editor(s). MDPI and/or the editor(s) disclaim responsibility for any injury to people or property resulting from any ideas, methods, instructions or products referred to in the content.

Proceeding Paper

A Review of the Effects of Project Management Practices on Cost Overrun in Construction Projects [†]

Fakhar Hassan Shah [1], Omer Shujat Bhatti [2] and Shehryar Ahmed [1,*]

[1] Department of Civil Engineering, Abasyn University Islamabad, Islamabad 45570, Pakistan; fakharshah15@gmail.com
[2] Shujat & Omer Architects, Islamabad 45570, Pakistan; omershujatbhatti@gmail.com
* Correspondence: engr.shehryar@outlook.com
[†] Presented at the 5th Conference on Sustainability in Civil Engineering (CSCE), Online, 3 August 2023.

Abstract: Cost overrun has long been a significant concern in the construction sector, posing obstacles to project profitability and financial viability. It occurs when actual costs exceed the initial budget estimates, leading to financial hardship, delays, and stakeholder disagreements. This paper investigates the impact of project management techniques on construction project cost overruns. The study aims to identify various techniques used at different stages of construction projects and analyze their effects on cost performance. Additionally, it explores the reasons behind cost overruns and proposes solutions to prevent them in the future. This research highlights efficient methods with which to manage and prevent cost overruns, providing valuable insights for project managers to use when improving cost performance and enhancing project success. It contributes to the knowledge on project management in the construction sector and aids stakeholders in navigating cost overrun challenges. Future research should explore context-sensitive issues related to cost overruns and consider robust, adaptive, and agile cost management strategies based on project management skillsets.

Keywords: cost overrun; project management practices; construction projects; remedial measures

1. Introduction

Project management practices are crucial for achieving successful construction projects, minimizing cost overruns, and ensuring positive outcomes. Key practices like efficient resource allocation, quality control, change management, and progress monitoring and control contribute to cost control and mitigate overruns [1]. Effective techniques such as scheduling, budget control, risk management, communication, collaboration, quality assurance, stakeholder engagement, and impact analysis enhance project efficiency and stakeholder satisfaction [2]. These practices are well documented in the literature, highlighting their positive impacts on project outcomes.

Cost overruns are a significant concern in the construction sector, hindering project profitability and financial viability. They happen when actual costs exceed the initial budget estimate, leading to financial hardship, delays, and stakeholder disagreements [3]. Studies have explored the relationship between project management practices and cost overruns, identifying accurate cost estimation, efficient scheduling, and resource allocation as crucial factors in prevention [4]. Moreover, project monitoring and control systems play vital roles in real-time tracking, the early identification of cost deviations, and the use of prompt corrective actions to prevent further escalation.

While existing studies have provided valuable insights, further research is essential to explore the specific effects of project management techniques on cost overruns in building projects. The connection between project management techniques and cost overrun in the construction sector contributes to the existing knowledge [5]. Effective project management techniques can significantly reduce the risk of cost overruns in building projects.

Key practices include accurate cost estimation, well-defined budgets, scope management, thorough project plans, proactive risk identification and mitigation, transparent stakeholder communication, and ongoing cost monitoring and control [3]. Consistent expense tracking, variance analyses, budget comparisons, and preventative actions are also crucial aspects to consider.

Overall, this research aims to provide valuable information to project managers, construction professionals, and other stakeholders. This was achieved through a thorough review of articles from reputable journals published in the last decade.

2. Project Management Practices in Different Stages of Construction Projects

Key project management practices are crucial for successful construction projects, minimizing cost overruns and ensuring positive outcomes. Comprehensive project planning involves detailed planning of scope, resources, timelines, and risks [1]. The use of project management practices at different project stages significantly impacts the achievement of a project's overall objectives by minimizing delays and improving efficiency, as shown in Figure 1. Adequate project management enables better budget control through accurate cost estimation, expense tracking, and cost management strategies [6]. Implementing quality control techniques like inspections, audits, and industry best practices leads to enhanced project deliverables and customer satisfaction [7]. The literature extensively documents the positive effects of these project management practices on construction projects, with benefits to outcomes, cost control, risk management, quality assurance, and stakeholder satisfaction.

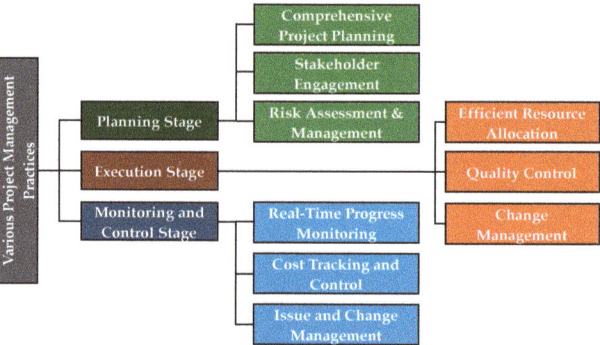

Figure 1. Project management practices.

3. Cost Overrun in Construction Projects

Cost overrun in construction projects arises from factors beyond the budget, necessitating effective project management and cost control to identify common causes [1]. Effective project management involves addressing these causes through proper estimation, scope management, and robust contract management to enhance performance. Table 1 presents studies related to remedial measures for cost overrun, emphasizing the significance of accurate cost estimation and budgeting as the foundations of effective project cost management [8]. Timely corrective actions are facilitated by effective cost control and monitoring, practices that identify cost deviations early [9]. Collaboration and communication promote cost-conscious decision making and shared understanding during construction projects [6]. Proactive risk management via risk assessments, response plans, and regular monitoring reduces cost overruns [8]. Tracking expenses, conducting variance analysis, and performing periodic audits are vital methods of effectively controlling and monitoring cost. Implementing remedial measures in construction projects, minimizing cost overruns, and improving performance necessitate a proactive approach, stakeholder commitment, and effective management practices.

Table 1. Remedial measures of cost overrun in construction projects.

Sr. no.	Remedial Measures of Cost Overrun	Impact	Ref.
1	Accurate Cost Estimation and Budgeting	Provides a solid foundation for managing project costs.	[1]
2	Effective Cost Control and Monitoring	Identify cost deviations and timely corrective actions.	[10]
3	Collaboration and Communication	Promotes cost-conscious decision making	[3]
4	Robust Project Planning and Scheduling	Enables cost forecasting for timely decision making.	[4]
5	Comprehensive Change Management	Change management reduces abrupt cost increases	[5]

4. Relationship between Project Management Practices and Cost Overrun

Effective project management practices significantly impact construction project cost overruns, minimizing risk and improving performance. A well-developed project plan with accurate cost estimation, realistic scheduling and sensible resource allocation helps prevent overruns. Table 2 shows the effects of project management practices on cost overrun in construction projects. Robust project planning and control enable timely corrective actions to avoid cost overruns [4]. Risk management minimizes the impact of risks on project costs and reduces the likelihood of overruns [11]. Stakeholder collaboration fosters cost-conscious decision making and reduces conflicts throughout the project lifecycle [7]. Procurement and contract management prevent cost overruns resulting from inflated prices, contractual disputes, or inadequate cost control by contractors [12]. Quality management minimizes cost-related issues arising from poor quality, leading to improved cost performance [6]. Implementing effective project management practices significantly re-duces the likelihood of cost overrun in construction projects [4]. The following section highlights key practices that can help prevent cost overruns. According to Figure 2, effective project management should inculcate integrated planning and scheduling with comprehensive scope management, effective cost estimation with monitoring and control, and effective communication with timely risk identification and mitigation.

Table 2. Effects of project management practices on cost overrun in construction projects.

Sr. no.	Project Management Practices	Effects/Results	Ref.
1	Project Planning and Control	Enable timely corrective actions to avoid cost overruns	[4]
2	Risk Management	Minimize impact of risks on project costs avoiding cost overruns	[11]
3	Stakeholder Collaboration	Reduce conflicts, and foster cost-conscious decision making	[7]
4	Contract Management	Prevent cost overruns from inflation and contractual disputes	[12]
5	Quality Management	Improving cost performance by avoiding poor quality work	[6]

Figure 2. Effective project management practices to avoid cost overrun in construction projects.

5. Conclusions

Cost overrun is a significant challenge in the construction industry, impacting project success and financial viability. Based on this review, it was evident that project management deployment and optimization acts as key factors in enabling cost management and the delivery of positive outcomes. The following conclusions can be drawn from the conducted study:

1. Effective cost control in construction requires crucial project management practices like accurate cost estimation, risk management, and resource allocation.
2. Real-time project monitoring and control systems play vital roles in enhancing cost control mechanisms, ensuring timely identification, and resolving cost overruns.
3. Integrating cost control mechanisms throughout the project lifecycle is essential for project success and financial viability in the face of cost overrun challenges.

Overall, this research provides valuable insights for project managers and stakeholders into the challenges of cost overrun and can be used to promote efficient project management practices.

6. Novelty and Future Research Direction

The research identified critical factors contributing to cost overrun in construction projects, with an influence of project management practices in managing costing issues. Further research is needed to collect context-sensitive data from construction sites in order to explore these variables in the local context and correlate them with their respective implications. The research added new dimensions to applicable project management aspects, although these may vary based on the site, nature, typology, scale, and size of construction projects.

Author Contributions: Conceptualization, S.A.; methodology, S.A. and O.S.B.; investigation, F.H.S.; data curation, F.H.S.; writing—original draft preparation, F.H.S.; writing—review and editing, S.A. and O.S.B.; supervision, S.A. and O.S.B. All authors have read and agreed to the published version of the manuscript.

Funding: This research received no external funding.

Institutional Review Board Statement: Not applicable.

Informed Consent Statement: Not applicable.

Data Availability Statement: Not applicable.

Acknowledgments: The authors would like to thank every person who supported in conducting this research. The careful review and constructive suggestions by the anonymous reviewers are gratefully acknowledged.

Conflicts of Interest: The authors declare no conflict of interest.

References

1. Ibrahim, A.H.; Elshwadfy, L.M. Factors affecting the accuracy of construction project cost estimation in Egypt. *Jordan J. Civ. Eng.* **2021**, *15*, 329–344.
2. Rehman, M.S.U.; Shafiq, M.T.; Ullah, F. Automated Computer Vision-Based Construction Progress Monitoring: A Systematic Review. *Buildings* **2022**, *12*, 1037. [CrossRef]
3. Sánchez-Garrido, A.J.; Navarro, I.J.; García, J.; Yepes, V. A systematic literature review on modern methods of construction in building: An integrated approach using machine learning. *J. Build. Eng.* **2023**, *73*, 106725. [CrossRef]
4. Khemakhem, M.A.; Chtourou, H. Efficient robustness measures for the resource-constrained project scheduling problem. *Int. J. Ind. Syst. Eng.* **2013**, *14*, 245. [CrossRef]
5. Yap, J.B.H.; Abdul-Rahman, H.; Chen, W. Collaborative model: Managing design changes with reusable project experiences through project learning and effective communication. *Int. J. Proj. Manag.* **2017**, *35*, 1253–1271. [CrossRef]
6. Aliyu, A.A.; Adamu, H.; Abdu, A.A.; Singhry, I.M. Influence of building contractors' performance on construction process in Nigeria: A review of emerging literature. *J. Energy Technol. Policy* **2015**, *5*, 11–22.
7. Carmona, S.; Ezzamel, M. Management Accounting and Strategy—A Review and Reflections on Future Research. *Eur. Account. Rev.* **2023**, 1–28. [CrossRef]

8. Ashkanani, S.; Franzoi, R. Gaps in megaproject management system literature: A systematic overview. *Eng. Constr. Arch. Manag.* **2022**, *30*, 1300–1318. [CrossRef]
9. Prater, J.; Kirytopoulos, K.; Ma, T. Optimism bias within the project management context: A systematic quantitative literature review. *Int. J. Manag. Proj. Bus.* **2017**, *10*, 370–385. [CrossRef]
10. Omar, T.; Nehdi, M.L. Data acquisition technologies for construction progress tracking. *Autom. Constr.* **2016**, *70*, 143–155. [CrossRef]
11. Afzal, F.; Yunfei, S.; Nazir, M.; Bhatti, S.M. A review of artificial intelligence based risk assessment methods for capturing complexity-risk interdependencies: Cost overrun in construction projects. *Int. J. Manag. Proj. Bus.* **2021**, *14*, 300–328. [CrossRef]
12. Habibi, M.; Kermanshachi, S. Phase-based analysis of key cost and schedule performance causes and preventive strategies: Research trends and implications. *Eng. Constr. Archit. Manag.* **2018**, *25*, 1009–1033. [CrossRef]

Disclaimer/Publisher's Note: The statements, opinions and data contained in all publications are solely those of the individual author(s) and contributor(s) and not of MDPI and/or the editor(s). MDPI and/or the editor(s) disclaim responsibility for any injury to people or property resulting from any ideas, methods, instructions or products referred to in the content.

Proceeding Paper

Project Management Practices in Construction Projects and Their Roles in Achieving Sustainability—A Comprehensive Review [†]

Fakhar Hassan Shah [1], Omer Shujat Bhatti [2] and Shehryar Ahmed [1,*]

1 Department of Civil Engineering, Abasyn University Islamabad, Islamabad 45570, Pakistan; fakharshah15@gmail.com
2 Shujat & Omer Architects, Islamabad 45570, Pakistan; omershujatbhatti@gmail.com
* Correspondence: engr.shehryar@outlook.com
† Presented at the 5th Conference on Sustainability in Civil Engineering (CSCE), Online, 3 August 2023.

Abstract: Effective project management practices are crucial in the construction sector, providing a structured approach to planning, executing, and controlling projects. They set clear objectives, define scopes, allocate resources efficiently, and manage risks effectively. However, challenges can arise throughout all project phases. This study focuses on literature from reputable journals over the last decade, and, considering the post-COVID scenario for inadequate scope definition, poor communication, resource mismanagement, and regulatory barriers were identified as major barriers to project success. To achieve sustainable construction projects, specific targets like energy efficiency, waste reduction, water conservation, and social responsibility must be set. Integrating project management with sustainability involves incorporating green building design, sustainable materials, waste management, water conservation, biodiversity promotion, smart technologies, and performance measurement systems. By adopting sustainable approaches and effective project management practices, construction projects can achieve successful outcomes while ensuring environmental responsibility, social equity, and economic viability. Future research should explore identified barriers, their local implications, and project management practices for successful project outcomes.

Keywords: construction projects; project management; sustainability targets; project phases

1. Introduction

Effective communication is crucial for project success, managing triple constraints, and sustainable practices. This involves environmental responsibility, social equity, and economic viability [1]. Project management methods control construction costs through precise estimation and cost control. Engaging stakeholders with effective communication ensure project success and quality assurance [2]. Proactive planning, stakeholder engagement, risk management, and continuous monitoring mitigate cost overruns, improving project progress for timely completion with minimized costs.

Energy efficiency, Leadership in Energy and Environmental Design (LEED) certifications [3], greenhouse gas (GHG) emissions reduction targets [4], and waste management are vital for sustainability [1]. Water conservation aims for efficient water management [1]. Challenges in achieving sustainability objectives include inadequate leadership support, limited resources, poor planning, and competing goals, as well as insufficient collaboration, communication, monitoring, and evaluation, and inconsistent regulations and policies creating barriers and uncertainties in reaching goals.

To achieve effective project planning and execution, define clear objectives, scope, deliverables, and success criteria. A detailed project plan and efficient communication with stakeholders prevent misunderstandings [1]. Implement risk management to maintain timelines and budgets [2]. Optimize resource management and adhere to quality standards [2]. Employ a structured change management process to handle changes. Monitor progress using KPIs while embracing innovative technologies, best practices, and green building principles for enhanced outcomes [2].

In view of deficient integration, adopting sustainable approaches, and implementing effective project management practices, construction projects can achieve successful outcomes while ensuring environmental responsibility, social equity, and economic viability.

2. Project Management Practices in Construction Sector

2.1. Role of Project Management in Construction Sector

Project management practices are crucial for successful construction projects, encompassing aspects like time, budget, quality, and resources. They prevent cost overrun issues through goal setting, feasibility studies, and realistic schedules [5]. Risk assessment and cost-control measures ensure staying within the budget [2]. In conclusion, project management is pivotal, providing a structured approach for planning, controlling, and executing, leading to successful project delivery.

2.2. Issues Related to Project Management during Various Project Phases

Project management involves distinct project lifecycle phases and challenges, as shown in Figure 1. Unclear initiation objectives and scope lead to stakeholder misalignment [2]. Inadequate feasibility assessment causes unrealistic goals and resource shortages [2]. Planning scope creep increases costs and delays [1]. Underestimating project duration and resource needs results in conflicts [1]. Proactive planning, communication, stakeholder engagement, risk management, and continuous monitoring are crucial for project success.

Figure 1. Issues related to project management during various project phases.

3. Sustainability in Construction Projects

3.1. Targets for Achieving Sustainability in Construction Projects

Achieving sustainability in construction requires specific targets for environmental, social, and economic responsibility. Table 1 lists common targets: energy efficiency to reduce consumption [6], (GHG) emissions reduction [4], waste management to minimize generation [1], water conservation [7], and sustainable materials use [8]. Realistic goals aligned with regulations, project type, and stakeholders' priorities contribute to a responsible built environment.

Table 1. Targets for achieving sustainability in construction projects.

Sustainability Factors	Targets	Ref.
Energy Efficiency	Establish energy performance targets in design to reduce energy consumption.	[6]
GHG Reduction	Set targets for carbon reduction by reducing GHG emissions.	[4]
Waste Management	Set waste minimization and recycling targets for construction wastes	[1]
Water Conservation	Set water consumption targets and implement technologies for management systems.	[7]
Sustainable Materials	Set targets for incorporating recycled and renewable materials in construction.	[8]

3.2. Issues Leading to Failure of Sustainability Goals

Challenges in building projects for sustainability include limited knowledge, poor planning, resource constraints, and competing goals. Leadership support, budget constraints, and rapid construction also pose difficulties. Inadequate monitoring hampers progress tracking [2]. Limited awareness, sustainable technology availability, and resistance to change hinder adoption [2]. Holistic sustainability demands strong leadership, stakeholder engagement, planning, communication, and resource allocation.

4. Integration of Project Management Practices with Sustainability

4.1. Effective Project Management Practices for a Successful Project

Effective project management practices are vital for successful construction projects, as shown in Figure 2. Key practices include well-defined objectives, scope, and deliverables, as well as effective communication with stakeholders [2]. Quality standards and a structured change management process are essential for meeting project standards. Systematic monitoring and reporting using KPIs is necessary [2]. A capable and motivated project team fosters a positive team culture. Embracing innovative technologies and industry best practices enhances project success and stakeholder satisfaction [9].

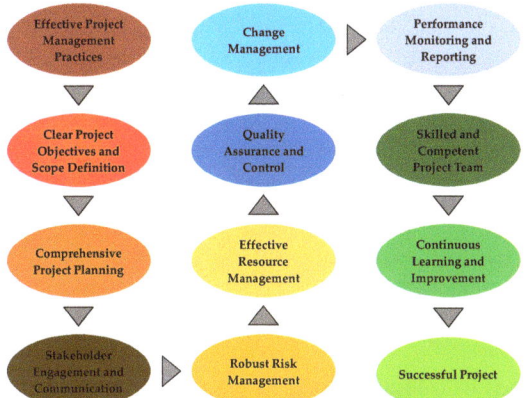

Figure 2. Effective project management practices to avoid cost overrun in construction projects.

4.2. Sustainable Approaches for a Successful Project

Table 2 shows achieving construction project sustainability through green building principles, using energy-efficient designs, sustainable materials, and renewable energy systems. Obtain certifications from recognized green building rating systems like LEED or other international standards [3]. Utilize water-efficient fixtures, rainwater harvesting, and greywater reuse [7]. Implement smart technologies like Building Management Systems (BMS) and IoT devices for energy efficiency and occupant comfort [9]. Set sustainability KPIs for energy, water, waste, and carbon emissions, regularly assessing and reporting progress for continuous improvement and transparency.

Table 2. Sustainable approaches for successful construction project.

Sustainable Approach	Practices to Be Adopted	Ref.
Water Conservation	Implement water-efficient fixtures and rainwater harvesting technologies	[7]
Green Design and Certification	Green building principles in design and certifications of rating systems	[3]
Biodiversity and Green Spaces	Sustainability by green spaces, urban forests, and biodiversity	[10]
Smart Technologies and BMS	Utilize BMS systems to optimize building efficiency and comfort	[9]
Performance check and Monitoring	Systems track water use and energy use to ensure sustainability	[2]

5. Conclusions

Project management ensures construction project success, aiding resource allocation, risk management, and communication. Integrating sustainability, green design, and smart technologies fosters environmental responsibility, social equity, and economic viability.

1. Project management and sustainability practices enhance success, cost control, stakeholder satisfaction, and built environment.
2. Sustainable construction projects aim for energy efficiency, waste reduction, water conservation, and social responsibility.
3. The blend of project management and sustainability fosters success, cost control, stakeholder satisfaction, and a sustainably built environment.

6. Recommendations

The research explores existing knowledge regarding practical aspects in construction projects, including training, deployment, standardization, and implications. Utilize key tools like scheduling, risk strategizing, quantitative analysis, and earned value management. Future research should focus on critical variables, variations, and reasons, guiding sustainable construction projects through project management skillset implications.

Author Contributions: Conceptualization, S.A.; methodology, S.A. and O.S.B.; investigation, F.H.S.; data curation, F.H.S.; writing—original draft preparation, F.H.S.; writing—review and editing, S.A. and O.S.B.; supervision, S.A. and O.S.B. All authors have read and agreed to the published version of the manuscript.

Funding: This research received no external funding.

Institutional Review Board Statement: Not applicable.

Informed Consent Statement: Not applicable.

Data Availability Statement: Not applicable.

Acknowledgments: The authors would like to thank every person who supported in conducting this research through accessibility to the information and the review of the literature, published data, and reports.

Conflicts of Interest: The authors declare no conflict of interest.

References

1. Borthakur, A.; Govind, M. Emerging trends in consumers' E-waste disposal behaviour and awareness: A worldwide overview with special focus on India. *Resour. Conserv. Recycl.* **2017**, *117*, 102–113. [CrossRef]
2. Lin, L.; Xu, F.; Ge, X.; Li, Y. Improving the sustainability of organic waste management practices in the food-energy-water nexus: A comparative review of anaerobic digestion and composting. *Renew. Sustain. Energy Rev.* **2018**, *89*, 151–167. [CrossRef]
3. Zhang, Y.; Wang, H.; Gao, W.; Wang, F.; Zhou, N.; Kammen, D.M.; Ying, X. A survey of the status and challenges of green building development in various countries. *Sustainability* **2019**, *11*, 5385. [CrossRef]
4. Mac Kinnon, M.A.; Brouwer, J.; Samuelsen, S. The role of natural gas and its infrastructure in mitigating greenhouse gas emissions, improving regional air quality, and renewable resource integration. *Prog. Energy Combust. Sci.* **2018**, *64*, 62–92. [CrossRef]
5. Govindaras, B.; Wern, T.S.; Kaur, S.; Haslin, I.A.; Ramasamy, R.K. Sustainable Environment to Prevent Burnout and Attrition in Project Management. *Sustainability* **2023**, *15*, 2364. [CrossRef]
6. Amiri, A.; Ottelin, J.; Sorvari, J. Are LEED-certified buildings energy-efficient in practice? *Sustainability* **2019**, *11*, 1672. [CrossRef]

7. Mugagga, F.; Nabaasa, B.B. The centrality of water resources to the realization of Sustainable Development Goals (SDG). A review of potentials and constraints on the African continent. *Int. Soil Water Conserv. Res.* **2016**, *4*, 215–223. [CrossRef]
8. Yaro, N.S.A.; Sutanto, M.H.; Baloo, L.; Habib, N.Z.; Usman, A.; Yousafzai, A.K.; Ahmad, A.; Birniwa, A.H.; Jagaba, A.H.; Noor, A. A Comprehensive Overview of the Utilization of Recycled Waste Materials and Technologies in Asphalt Pavements: Towards Environmental and Sustainable Low-Carbon Roads. *Processes* **2023**, *11*, 2095. [CrossRef]
9. Nižetić, S.; Djilali, N.; Papadopoulos, A.; Rodrigues, J.J. Smart technologies for promotion of energy efficiency, utilization of sustainable resources and waste management. *J. Clean. Prod.* **2019**, *231*, 565–591. [CrossRef]
10. Sundara Rajoo, K.; Karam, D.S.; Abdu, A.; Rosli, Z.; James Gerusu, G. Urban forest research in Malaysia: A systematic review. *Forests* **2021**, *12*, 903. [CrossRef]

Disclaimer/Publisher's Note: The statements, opinions and data contained in all publications are solely those of the individual author(s) and contributor(s) and not of MDPI and/or the editor(s). MDPI and/or the editor(s) disclaim responsibility for any injury to people or property resulting from any ideas, methods, instructions or products referred to in the content.

Proceeding Paper

Effectiveness of Mono Sand Piles in Soft Cohesive Ground [†]

Qazi Umar Farooq * and Muhammad Tayyab Naqash

Department of Civil Engineering, Islamic University of Madinah, Al-Madinah al-Munawwarah 42351, Saudi Arabia; engr.tayyabnaqash@gmail.com
* Correspondence: umar@iu.edu.sa
† Presented at the 5th Conference on Sustainability in Civil Engineering (CSCE), Online, 3 August 2023.

Abstract: Soft cohesive formations are extensively distributed across the earth's land mass. They mainly comprise medium to high plastic clays deposited by thousands of years of geological activity. In Pakistan, the upper and lower plains of the Indus Valley have several square kilometers of cohesive ground. The cohesive soils are vulnerable to moisture variations and lack friction. Hence, they are not considered an ideal ground for foundation support. The raft foundation and traditional reinforced concrete piles are effective solutions, but are uneconomical. Sand piles can replace these expansive foundations for moderately loaded structures; however, their effectiveness is required to be supported by field and research investigations. This study presents FEM-based numerical investigations on the performance of a single sand pile on soft cohesive ground. The pile is loaded with the 100 kPa pressure, representing a moderately loaded structure. The stress–strain behaviors and overall pile settlement results are graphically presented. The sand pile, the stiffer material, could hold most of the stresses while maintaining volumetric strains up to 10%, thus allowing better load transfer to the naturally soft ground.

Keywords: cohesive soil; foundation settlement; numerical modeling; sand pile; stress distribution

1. Introduction

Fine-grained soils (less than 0.075 mm in size) are problematic when they comprise medium to high plastic clays. The fertile agricultural plains of Punjab and Sindh mainly consist of cohesive soils formed by the alluvial deposits of the river Indus and its tributaries. Cohesive soils have threaded particles that lack friction but have reasonable cohesion in a dry state; however, under moist conditions, they lose their strength and become soft. Therefore, they are considered inadequate for sustaining foundations [1]. Traditional solutions like raft foundations and reinforced concrete piles function well. However, they might not be economically viable, especially for moderately loaded structures. On the other hand, conventional RC piles sometimes require unique and complex driving techniques [2]. Sand piles have been proposed as a potential replacement for conventional deep foundations up to a certain degree. Sand piling is a ground-improvement technique that replaces the inadequate soil layer with sand piles produced by drilling holes into the ground and filling them with dense sand [3]. These piles are used to construct a sequence of columns that support the foundation. Since they can reduce settlement and increase the bearing capacity of the soft ground, sand piles are a possible replacement for traditional deep foundations. The performance of sand piles can be affected by several variables, including the pile's size and shape, the soil's characteristics, and the load circumstances. Therefore, numerical modeling is a valuable tool for analyzing the performance of sand piles under various conditions [4]. Nevertheless, sand piles cannot be applied to structures subjected to complex loading and specific drainage requirements [5]. Numerical simulations utilizing various codes, such as PLAXIS and COMSOL [6], have become an effective tool for researching the behavior of soils under various loading circumstances; these models can assist engineers in designing more stable and dependable soil structures. Recent studies have stressed the importance of taking time-dependent behavior into account when evaluating the stability and performance of soil by utilizing numerical simulations [7].

Research investigations to assess the efficacy of sand piles in soft cohesive ground are limited; an effort has been made in this brief study to analyze stress–strain behaviors and overall sand pile settlement. An FEM-based numerical model using COMSOL Multiphysics code has been used in this perspective. This will give readers an insight into how sand piles perform under static loads.

2. Materials and Methods

The natural ground was represented by fine-grained soil with a soft consistency, whilst dense sand simulated the sand pile. The fundamental properties of geomaterials were determined in the laboratory, and elastic properties were estimated using classical correlations [6]. The mechanical properties of both components of the sand pile system are shown in Table 1.

Table 1. Soil characteristics of natural ground and the sand pile.

System Component	Unit Weight (Y) kN/m³	Soil Friction Angle (Φ)	Soil Cohesion (C) kPa	Elastic Modulus (E) MPa	Poisson's Ratio (n)
Natural Ground	14.71	-	50	40	0.3
Sand Pile	19.60	32	-	75	0.25

The 2-D model was developed to analyze the 6 m long sand pile embedded in a 12 m deep natural ground comprising soft cohesive soil. The analysis was carried out using the COMSOL Multiphysics program's geomechanics module. The model geometry's main features and boundary conditions required to precisely reproduce pile behavior can be seen in Figure 1. A moderately loaded structure was simulated using 100 kPa applied stress.

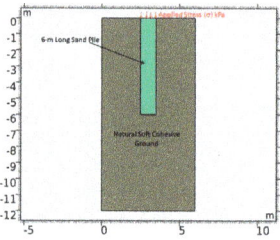

Figure 1. Cross-section of the sand pile and boundary conditions.

3. Results and Discussion

The sand pile stress absorption was examined through the resultant stress diagram. Figure 2a depicts the stress variation in the sand pile system, demonstrating the lower stress distribution in the natural ground compared with that in the sand pile. The stress distribution around the sand pile is focused on the applied stress contact point, demonstrating that the pile holds most of the stresses and facilitates better load transfer to the natural ground. Since the sand pile is stiffer than the surrounding soft soil, there is stress concentration around the pile cap. The sand pile stress distribution and soil displacement behavior are, respectively, illustrated in Figure 2a,b. The soil displacement behavior is a key performance indicator of the sand pile in soft cohesive soils. The analytical results show slight displacement within the sand pile system under the specified load settings. Compared with the nearby soil, the displacement is predominantly centered around the pile and is relatively minimal. The increased stiffness of the sand pile, which prevents the surrounding soil from deforming, is the reason for the sand pile system's displacement behavior.

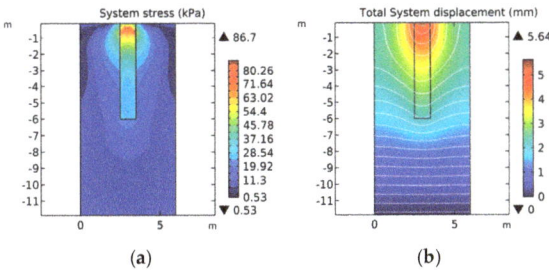

Figure 2. Stress-displacement behavior of the sand pile system. (**a**) Stress Distribution, (**b**) Soil Displacement.

The stress–strain range at the cap and tip of the pile is shown in Figure 3. The analysis's findings show the pile's ability to support loads and its distortion. The volumetric strain gauges the volume change in the soil surrounding the pile. The pile's cap and tip's negative values for volumetric strain indicate soil compression around the pile; the volumetric strain is more significant at the pile cap than the pile tip.

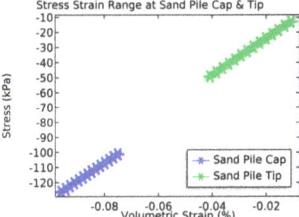

Figure 3. Stress–strain range at pile extremes.

The sand pile was subjected to a static load. However, the consolidation phenomena in cohesive material affected the overall performance. The settlement rate was measured using the system displacement measured at the start, i.e., 0 min, and at 30 min and 60 min intervals. Figure 4a demonstrates how the applied load deforms the sand pile. The displacement of the pile increases over time, reaching its maximum displacement at 60 min. This implies that the pile deforms more with time, proving that a time-dependent deformation under a sustained load is occurring.

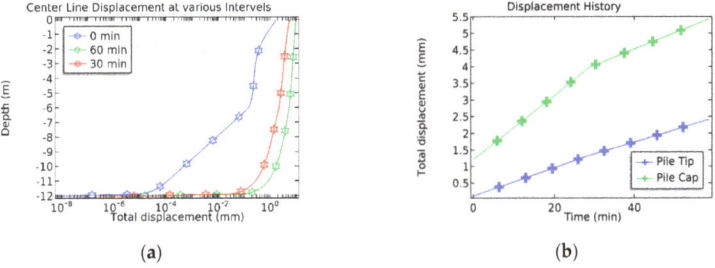

Figure 4. Consolidation effects of the sand pile system. (**a**) Depth (m) vs. displacement (mm), (**b**) Displacement history.

It is also important to note that the deformation rate changes over time. At the start of the simulation, the deformation rate is the fastest. As time goes on, the displacement increases at a slower phase (see Figure 4b). This non-linear behavior is typical for soils since soils display complicated stress–strain behavior based on the type, moisture content, and applied load. Overall, the results shed light on the sand pile's time-dependent behavior and emphasize the significance of considering it for the stability of structures.

4. Conclusions

The following conclusions can be drawn from the presented results.

- Numerical simulations can be economically used to predict the mechanical behavior of complex soil systems such as sand piles in soft cohesive ground.
- The sand pile, a stiffer material, can bear most of the imposed stresses while reflecting low strain values, resulting in overall low settlements of the foundation system.
- The time-dependent or consolidation behavior of the foundation remains a challenge in cohesive ground even after the installation of the sand pile.

The research outcomes can be a benchmark for analyzing the substructures required to be built in complex soil conditions.

Author Contributions: Conceptualization, Q.U.F.; methodology, Q.U.F. and M.T.N.; software, M.T.N.; validation, Q.U.F. and M.T.N.; formal analysis, Q.U.F.; investigation, Q.U.F.; resources, M.T.N.; data curation, M.T.N.; writing—original draft preparation, Q.U.F.; writing—review and editing, Q.U.F. and M.T.N.; visualization, Q.U.F. and M.T.N.; supervision, Q.U.F.; All authors have read and agreed to the published version of the manuscript.

Funding: This research received no external funding.

Institutional Review Board Statement: Not applicable.

Informed Consent Statement: Not applicable.

Data Availability Statement: The laboratory test results, figures, and tables used to support the findings of this study are included within the article.

Conflicts of Interest: The authors declare no conflict of interest.

References

1. Naqash, M.T.; Farooq, Q.U. Performance of Rigid Steel Frames under Adequate Soil Conditions Using Seismic Code Provisions. *Open J. Civ. Eng.* **2018**, *8*, 91–101. [CrossRef]
2. Kementzetzidis, E.; Pisanò, F.; Elkadi, A.S.K.; Tsouvalas, A.; Metrikine, A.V. Gentle Driving of Piles (GDP) at a sandy site combining axial and torsional vibrations: Part II—cyclic/dynamic lateral loading tests. *Ocean Eng.* **2023**, *270*, 113452. [CrossRef]
3. Hussein, M. Effect of Sand and Sand-Lime Piles on the Behavior of Expansive Clay Soil. *Adv. Civ. Eng.* **2021**, *2021*, 4927078. [CrossRef]
4. Harireche, O.; Naqash, M.T.; Farooq, Q.U. A full numerical model for the installation analysis of suction caissons in sand. *Ocean Eng.* **2021**, *234*, 109173. [CrossRef]
5. Chen, Z.; Wang, B.; Wang, C.; Wang, Y.; Xiao, P.; Li, K. Performance of a Subgrade-Embankment-Seawall System Reinforced by Drainage PCC Piles and Ordinary Piles Subjected to Lateral Spreading. *Geofluids* **2023**, *2023*, 4489478. [CrossRef]
6. Farooq, Q.U.; Naqash, M.T. Performance of Shallow Building Foundations under Infrequent Rainfall Patterns at Al-Madinah, Saudi Arabia. *Open Civ. Eng. J.* **2021**, *15*, 91–103. [CrossRef]
7. Al-Ramthan, A.Q.O.; Aubeny, C.P. Numerical Investigation of the Performance of Caissons in Cohesive Soils under Cyclic Loading. *Int. J. Geomech.* **2020**, *20*, 04020042. [CrossRef]

Disclaimer/Publisher's Note: The statements, opinions and data contained in all publications are solely those of the individual author(s) and contributor(s) and not of MDPI and/or the editor(s). MDPI and/or the editor(s) disclaim responsibility for any injury to people or property resulting from any ideas, methods, instructions or products referred to in the content.

Proceeding Paper

Effect of Bio-Char of Santa Maria Feverfew Plant on Physical Properties of Fresh Mortar [†]

Waleed Nasir Khan, Syed Ghayyoor Hussain Kazmi and Anwar Khitab *

Department of Civil Engineering, Mirpur University of Science and Technology (MUST), Mirpur 10250, Pakistan; khanwaleed1840@gmail.com (W.N.K.); kazmighayyoor4156@gmail.com (S.G.H.K.)
* Correspondence: anwar.ce@must.edu.pk
† Presented at the 5th Conference on Sustainability in Civil Engineering (CSCE), Online, 3 August 2023.

Abstract: The present study concerns the application of nano-/micro-sized fibers (bio-char of Santa Maria feverfew) in cementitious mortars. The bio-char was added @ 0, 0.05 and 0.1% by mass of cement. The addition of bio-char did not affect the setting and consistency of the mortars. The fresh density was reduced by 11%, while the followability decreased by 50%. It is concluded that the bio-char results in a light-weight cementitious material, without affecting the setting time or consistency. Bio-char produces carbon-rich materials, the use of which as building materials adds to carbon sequestration in accordance with the Sustainable Development Goals of the UNO.

Keywords: bio-char; Santa Maria feverfew plant; cementitious mortar; fresh density; consistency; flowability; carbon sequestration; sustainable development goal

1. Introduction

Pyrolysis is an endoergic method that encompasses the thermo-chemical breakdown of raw bio-mass in an inert atmosphere at high temperatures and pressures. This procedure yields different valuable products like bio-char, liquid bio-oil and fuel gases [1]. Bio-char is a light-weight, dark-colored carbon deposit. Scientists have used many different kinds of feedstock, including water hyacinth, oriental beech, corncob and many more [2–4]. In recent years, many researchers have added bio-char to cementitious products to enhance their performance. Gupta et al. added bio-char of sawdust at a rate 2% by mass of cement [5]. The results revealed that the addition enhanced the compressive strength and ductility of the end products. Tayyab et al. incorporated the bio-char of millet and maize in mortar [6]. The authors reported an enhancement of the fracture toughness and ductility of the specimens. This enhancement was attributed to crack bridging/branching due to the fibrous nature of the bio-char. Iftekhar et al. studied the effect of bio-chars of sugarcane bagasse and pine needles in cementitious mortars [7]. The authors reported enhanced interface shielding. Restuccia et al. used the bio-char of hazelnut shells in mortars. They reported an enhanced compressive strength, flexural strength, toughness and ductility. Ling et al. used bio-char @ 1–3% by mass of cement in mortar. Their findings revealed that a 3% bio-char content enhanced the degree of hydration [8]. Mensah et al. reported that bio-char has the capacity to enhance the mechanical and thermal properties of cementitious materials [9]. Most of the previous studies focused on the effects of bio-char on the hardened properties of the cementitious composites. Literature regarding the influence of bio-char addition on the fresh properties of cementitious materials is scarce.

The present study focusses on the addition of the bio-char of Santa Maria feverfew on the fresh characteristics of cementitious mortar. Santa Maria feverfew is a local plant known as gajjar boti. The use of the Santa Maria plant for producing bio-char and that of its bio-char as filler in mortar has not been studied previously. This study comprises of an evaluation of the cementitious mortar in terms of its fresh density, consistency, setting time and flowability.

Citation: Nasir Khan, W.; Kazmi, S.G.H.; Khitab, A. Effect of Bio-Char of Santa Maria Feverfew Plant on Physical Properties of Fresh Mortar. *Eng. Proc.* **2023**, *44*, 4. https://doi.org/10.3390/engproc2023044004

Academic Editors: Majid Ali, Muhammad Ashraf Javid, Shaheed Ullah and Iqbal Ahmad

Published: 22 August 2023

Copyright: © 2023 by the authors. Licensee MDPI, Basel, Switzerland. This article is an open access article distributed under the terms and conditions of the Creative Commons Attribution (CC BY) license (https://creativecommons.org/licenses/by/4.0/).

2. Materials and Methods

Cementitious mortars containing both the control and the specimens with added bio-char were prepared. A C-53-grade local ordinary Portland cement was used (Table 1). River sand was used as a fine aggregate (Table 1). The transformation of the Santa Maria Feverfew plant from a raw product to pyroletic powder is shown in Figure 1. Three types of specimens were prepared: a control and those containing 0.05% and 0.1% bio-char as the mass of cement. The composition of the materials is presented in Table 2.

Table 1. Chemical and physical characteristics of the ingredients of the mortar.

Cement				Sand	
Compound	%Age	Property	Value	Property	Value
CaO	61	Specific gravity	3.1	Specific gravity	2.7
SiO_2	21	Soundness	2%	Fineness modulus	2.7
Fe_2O_3	3	Fineness	1%	Bulk density (Kg/m^3)	1480
Al_2O_3	6	Initial setting time	30 min	Dry-rodded bulk density(Kg/m^3)	1820
MgO	1.5	Final setting time	610 min	Water absorption(%)	3.9
Alkalis	0.5	Consistency	24%	Water Content (%)	1.98
Gypsum	4				

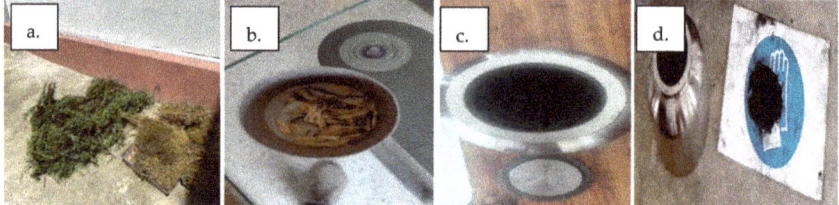

Figure 1. (a) Fresh Santa Maria Feverfew plant. (b) Dry plant. (c) Bio-char. (d) Powdered bio-char.

Table 2. Composition of the mortar.

Samples	OPC (g)	Sand (g)	Water (mL)	Bio-Char (g)	Admixture (mL)	W/C
C0	610	916	214	0	6	0.35
C0.05	610	916	214	0.305	6	0.35
C0.1	610	916	214	0.61	6	0.35

The materials were mixed as per the ASTM C305-20 standard method [10]. To avoid agglomeration, the bio-char was mixed in water via the UV–sonication technique. An admixture (super-plasticizer) was added (1% by mass of cement) to make the dispersion more effective. All the ingredients were mixed in a Hobart mixer. After mixing, the samples were cast in prisms (40 × 40 × 160 mm) according to the ASTM C1314 method [11]. After 24 h, the specimens were de-molded and immersed in water for curing. The flowability was measured through a flow table test (ASTM C1437) [12]. The consistency of the mix was determined using a Vicat apparatus (ASTM C187) [13]. The fresh density of the specimens was determined using the ASTM C138/138M method [14]. The mixing machine, molding and flowability test are shown in Figure 2.

Figure 2. (a) Mixing, (b) molding and (c) flowability test.

3. Results and Discussion

3.1. Setting Time

The initial setting times are presented in Table 3. The results show that there is a slight increase in the setting time, yet it is well within the limit, i.e., 30–45 min. The results show that the addition of 0.05–0.1% of the bio-char of Santa Maria Feverfew does not affect the initial setting time; the slight enhancement may be the result of the porous nature of the particles, which absorb water, thus delaying hydration. The final setting times are presented in Table 3. There is an almost negligible effect on the final setting time. Hence, the bio-char particles do not affect the setting time of the paste. The filler particles absorb water and, as such, may affect the hydration as well as the setting time, which is dependent on the water [15].

Table 3. Effects of bio-char on the setting time and consistency of cement paste.

Specimen	Initial Setting Time (min)	Final Setting Time (min)	Consistency (%)
C0	33.48	495	23.5
C0.05	35.32	485	23.7
C0.1	39.30	500	24.0

3.2. Consistency

The effect of the bio-char on the consistency of the cement paste is shown in Table 3. Consistency is an important measure, as it provides an indication of the amount of water for cement hydration. The slight enhancement may be the result of the porous nature of the bio-char. Since bio-chars are porous, they may absorb water and enhance the water requirements, but again, this enhancement is within the limits (25–30%).

3.3. Fresh Density

The fresh density of the mortar specimens is shown in Table 4. The results show that the density reduces with the addition of the bio-char.

Table 4. Effect of bio-char on fresh density.

Specimen	Density (kg/m^3)	% Difference
C0	2100	
C0.05	2000	4.8
C0.1	1880	10.5

The addition of bio-char enhances the volume of the mix. Since bio-chars are extremely light materials, the increase in mass is small as compared to the increase in volume. This, in turn, reduces the density. These results are in accordance with the previous literature [6,7].

3.4. Flowability

Flowability was assessed through a flow table test. The results are shown in Table 5.

Table 5. Effect of bio-char on flowability.

Specimen	D1 (mm)	D2 (mm)	D3 (mm)	Mean D (mm)	% Difference
C0	9.7	10.1	10.2	10	
C0.05	5.1	5.4	5.7	5.4	46
C0.1	4.9	4.8	4	4.6	54

The results show that the flowability drastically decreases with the addition of bio-char. The absorption of moisture by the bio-char particles reduces the flow. Bio-char particles are finer than cement. Finer particles enhance the specific area of the particles and, hence, lessen the amount of water for the lubrication of the mix. Therefore, there is a substantial reduction in flow. This situation demands the use of a larger amount of superplasticizer to maintain the flow. In the present study, the superplasticizer was mainly introduced for the dispersion of the bio-char particles. Owing to their nano-size, the particles become agglomerated.

The use of bio-char for the enhancement of the properties of cementitious products in a hardened form has been documented by many researchers. The present work explored the effects of bio-char on the fresh properties of cementitious materials. The results revealed that the bio-chars can be used as admixtures for reducing fresh density, with no effect on the setting time.

Author Contributions: Conceptualization, A.K.; methodology, W.N.K. and S.G.H.K.; validation, W.N.K. and S.G.H.K.; formal analysis, W.N.K. and S.G.H.K.; investigation, W.N.K. and S.G.H.K.; resources, A.K.; data curation, W.N.K.; writing—original draft preparation, A.K.; writing—review and editing, A.K.; visualization, S.G.H.K.; supervision, A.K.; project administration, A.K.; funding acquisition, A.K. All authors have read and agreed to the published version of the manuscript.

Funding: This research received external funding from Higher Education Commission of Pakistan through NRPU-7984 project.

Institutional Review Board Statement: Not applicable.

Informed Consent Statement: Not applicable.

Data Availability Statement: The authors confirm that the data supporting the findings of this study are available within the article.

Conflicts of Interest: The authors declare no conflict of interest.

References

1. Khitab, A.; Ahmad, S.; Khan, R.A.; Arshad, M.T.; Anwar, W.; Tariq, J.; Khan, A.S.R.; Khan, R.B.N.; Jalil, A.; Tariq, Z. Production of Biochar and Its Potential Application in Cementitious Composites. *Crystals* **2021**, *11*, 527. [CrossRef]
2. Carnaje, N.P.; Talagon, R.B.; Peralta, J.P.; Shah, K.; Paz-Ferreiro, J. Development and characterisation of charcoal briquettes from water hyacinth (*Eichhornia crassipes*)-molasses blend. *PLoS ONE* **2018**, *13*, e0207135. [CrossRef]
3. Aburas, H.; Demirbas, A. Evaluation of beech for production of bio-char, bio-oil and gaseous materials. *Process Saf. Environ. Prot.* **2015**, *94*, 29–36. [CrossRef]
4. Gupta, G.K.; Ram, M.; Bala, R.; Kapur, M.; Mondal, M.K. Pyrolysis of chemically treated corncob for biochar production and its application in Cr(VI) removal. *Environ. Prog. Sustain. Energy* **2018**, *37*, 1606–1617. [CrossRef]
5. Gupta, S.; Kua, H.W.; Low, C.Y. Use of biochar as carbon sequestering additive in cement mortar. *Cem. Concr. Compos.* **2018**, *87*, 110–129. [CrossRef]
6. Tayyab, S.; Khitab, A.; Iftikhar, A.; Khan, R.B.N.; Kirgiz, M.S. Manufacturing of high-performance light-weight mortar through addition of biochars of millet and maize. *Waste Dispos. Sustain. Energy* **2023**, *5*, 97–111. [CrossRef]
7. Iftikhar, A.; Arsalan Khushnood, R.; Khitab, A.; Ahmad, S. Improved fracture response and electromagnetic interference shielding effectiveness of cementitious composites incorporating pyrolytic bagasse fibers and pine needles. *Constr. Build. Mater.* **2023**, *370*, 130722. [CrossRef]
8. Ling, Y.; Wu, X.; Tan, K.; Zou, Z. Effect of Biochar Dosage and Fineness on the Mechanical Properties and Durability of Concrete. *Materials* **2023**, *16*, 2809. [CrossRef] [PubMed]
9. Mensah, R.; Shanmugam, V.; Narayanan, S.; Razavi, N.; Ulfberg, A.; Blanksvärd, T.; Sayahi, F.; Simonsson, P.; Reinke, B.; Försth, M.; et al. Biochar-Added Cementitious Materials—A Review on Mechanical, Thermal, and Environmental Properties. *Sustainability* **2021**, *13*, 9336. [CrossRef]

10. *ASTM C305*; Standard Practice for Mechanical Mixing of Hydraulic Cement Pastes and Mortars of Plastic Consistency. ASTM International: West Conshohocken, PA, USA, 2022.
11. *ASTM C1314-21*; Standard Test Method for Compressive Strength of Masonry Prisms. ASTM International: West Conshohocken, PA, USA, 2021.
12. *A14STM C37-20*; Standard Test Method for Flow of Hydraulic Cement Mortar. ASTM International: West Conshohocken, PA, USA, 2020.
13. *ASTM C187-16*; Standard Test Method for Amount of Water Required for Normal Consistency of Hydraulic Cement Paste. ASTM International: West Conshohocken, PA, USA, 2015.
14. *ASTM C138/C138M*; Standard Test Method for Density (Unit Weight), Yield, and Air Content (Gravimetric) of Concrete. ASTM International: West Conshohocken, PA, USA, 2017.
15. Elyamany, H.E.; Abd Elmoaty, A.E.M.; Mohamed, B. Effect of filler types on physical, mechanical and microstructure of self compacting concrete and Flow-able concrete. *Alex. Eng. J.* **2014**, *53*, 295–307. [CrossRef]

Disclaimer/Publisher's Note: The statements, opinions and data contained in all publications are solely those of the individual author(s) and contributor(s) and not of MDPI and/or the editor(s). MDPI and/or the editor(s) disclaim responsibility for any injury to people or property resulting from any ideas, methods, instructions or products referred to in the content.

Proceeding Paper

Linear and Non-Linear Regression Analysis on the Prediction of Compressive Strength of Sodium Hydroxide Pre-Treated Crumb Rubber Concrete [†]

Hamza Aamir *, Kinza Aamir and Muhammad Faisal Javed

Department of Civil Engineering, COMSATS University Islamabad, Abbottabad Campus, Abbottabad 22060, Pakistan; kinzamuhammadaamir@gmail.com (K.A.); arbabfaisal@cuiatd.edu.pk (M.F.J.)
* Correspondence: hamzaamir9696@gmail.com
† Presented at the 5th Conference on Sustainability in Civil Engineering (CSCE), Online, 3 August 2023.

Abstract: This research focuses on dataset development using NaOH treatment period (NaTP), NaOH concentration (NaCon), coarse aggregates (gravel), fine aggregates (sand), water, water–cement ratio (w/c), crumb rubber percentage (CR%), and equations to predict the CS of concrete. The criteria for the model accuracy included the coefficient of regression (R^2), mean absolute error (MAE), and root mean square deviation (RMSE). In this study, Multiple Non-Linear Regression (MNLR) performed better compared to Multiple Linear Regression (MLR). The MNLR values obtained for R^2, MAE, and RMSE were 0.88, 4.64, and 6.15; and the MLR values were 0.82, 5.86, and 7.43 for R^2, MAE, and RMSE, respectively.

Keywords: regression analysis; multiple linear regression; pre-treatment; compressive strength

1. Introduction

Concrete structures play an essential role in providing shelter, housing, transportation, and various aspects of construction [1]. Concrete is made up of fine aggregate, coarse aggregate, and cement when mixed with water [2]. Aggregates in the construction and mining processes are depleting natural resources [2]. The current scenario faces a major problem with industrial waste, posing a threat to the environment. Researchers use waste products to create sustainable cementitious composites to address these major problems. [3]. Waste in excessive amounts contributes to pollution, which in turn is harmful to living habitats [4–6].

A recent study used 152 datasets to forecast 28 days of compressive strength of high-performance concrete with metakaolin. The models used were Linear Regression (LR), Multi-Logistic Regression (MLR), Response Surface Methodology (RSM), and Non-Linear Regression (NLR). The RSM model performed best, providing results close to those of laboratory testing. The sequence of accuracy was RSM > NLR > LR > MLR [7]. A study on the compressive strength of geopolymer mortar using statistical predictive models was performed using 247 datasets. The models used were LR, MLR, and NLR. The NLR outperformed the other two models in forecasting and real-time analysis, indicating a greater reliance on NLR [8].

This research focuses on predicting the compressive strength of NaOH pre-treated crumb rubber concrete using statistical models with basic evaluation criteria for the accurate forecasting of CS. This approach saves time, reduces laboratory testing, and is convenient for designers.

2. Research Procedure

2.1. Dataset Development

The dataset for the linear and non-linear regressions was obtained from international journals [9–17]. In this dataset, different parameters were used to create a linear and non-linear model in order to avoid the hectic task of casting samples. The dataset included the NaOH-treated crumb rubber data with 115 entries. The input variables used in the dataset were the NaOH treatment period (NaTP), NaOH concentration (NaCon), crumb rubber percentage (CR%), water-to-cement ratio (w/c), cement, water, fine aggregates (sand), and coarse aggregates (gravel). NaTP is the period during which the crumb rubber is placed in the prepared solution of sodium hydroxide and water to make its surface rough for better bonding, NaCon is the amount of sodium hydroxide in the water for the preparation of the solution: 10% of NaCon indicates that 10% sodium hydroxide was added to water to prepare the solution as the treatment solution, and CR% is the percentage of the tire rubber, which is shredded and to be used as a replacement for sand in the mix. The output parameter used in the dataset was the compressive strength of NaOH-treated crumb rubber after 28 days of curing (Table 1).

Table 1. Input and output parameters.

Parameters	NaTP	NaCon	%CR	w/c	Cement	Water	Sand	Gravel	CS
Mean	7.40	8.07	10.00	0.45	377.55	173.92	699.46	607.81	36.41
Standard Error	1.01	0.81	0.75	0.01	8.59	4.45	20.94	36.28	1.21
Median	0.50	10.00	10.00	0.48	396.00	175.00	685.00	416.00	36.72
Mode	0.00	10.00	0.00	0.50	400.00	175.00	990.00	311.00	35.00
Standard Deviation	10.81	8.64	8.08	0.06	92.11	47.74	224.58	389.09	12.96
Sample Variance	116.85	74.59	65.25	0.00	8483.41	2279.06	50,437.86	151,393.70	167.99
Kurtosis	−1.20	4.85	−0.84	−0.25	8.03	5.01	2.42	−1.52	−0.18
Skewness	0.90	1.77	0.27	−0.36	−2.35	−1.23	−0.24	0.15	−0.29
Range	24.00	50.00	30.00	0.25	481.20	284.60	1333.20	1257.00	61.30
Minimum	0.00	0.00	0.00	0.35	18.80	9.20	41.80	0.00	1.70
Maximum	24.00	50.00	30.00	0.60	500.00	293.80	1375.00	1257.00	63.00
Sum	851.34	928.00	1150.00	52.29	43,418.00	20,001.09	80,438.20	69,898.40	4187.19
Count	115.00	115.00	115.00	115.00	115.00	115.00	115.00	115.00	115.00

2.2. Error Evaluation

After the model development, the accuracy of the model depends on the error, which can be calculated with the help of statistics. Each error has different criteria to check the accuracy of the model. In this model, the first error evaluation method that was used was the R^2 coefficient of determination; its value normally ranges between 0 and 1, and a value close to 1 indicates that the error is less. 'A' represents the observed or actual value while 'F' represents the forecasted value.

$$R^2 = \frac{\sum_{i=1}^{n}(A_i - \overline{A}_i)(F_i - \overline{F}_i)}{\sqrt{\sum_{i=1}^{n}(A_i - \overline{A}_i)^2 \sum_{j=1}^{p}(F_i - \overline{F}_i)^2}} \quad (1)$$

The second error valuation criterion used in this research was the root mean square deviation (RMSE). In the case of RMSE, a value closer to 0 is good as compare to a value farther from 0, so 0 is considered the benchmark. In the equation of RMSE, the actual value is represented by 'A' and the forecasted value by 'F'.

$$\text{RMSE} = \sqrt{\frac{\sum_{i=1}^{n}(A_i - F_i)^2}{n}} \qquad (2)$$

The mean absolute error was also used in this research; the closer the value to zero, the better the results predicted.

$$\text{MAE} = \frac{1}{n}\sum_{i=1}^{n}|F_i - A_i| \qquad (3)$$

3. Results

3.1. MLR Model

The general representation of linear and non-linear regression is represented in Figure 1. With the help of the input parameters, i.e., NaTP, NaCon, CR%, w/c, water, sand, and gravel, the equation was developed for the MLR model in an Excel data sheet for forecasting and results extraction for the dependent variable. The equation developed for the prediction of NaOH-pre-treated crumb rubber is shown in Equation (4).

$$CS = -28.966 - 0.571\text{NaTP} + 0.136\text{NaCon} - 0.564\%\text{CR} + 120.039\text{w/c} + 0.303\text{Cement} - 0.496\text{Water} + 0.006\text{Sand} - 0.021\text{Gravel} \qquad (4)$$

Figure 1. General figure for (a) linear and (b) non-linear regression.

3.2. MNLR Model

Independent variables, i.e., NaTP, NaCon, CR%, w/c, water, sand, and gravel, were used for the development of an equation for forecasting the compressive strength of NaOH-pre-treated crumb rubber concrete. The equation developed using the Excel dataset for the prediction is shown in Equation (5).

$$CS = 77.540 - 2.795\text{NaTP} + 0.093\text{NaTP}^2 + 1.239\text{NaCon} - 0.025\text{NaCon}^2 - 1.395\%\text{CR} + 0.019\%\text{CR}^2 - 243.610\text{w/c} + 253.984\text{w/c}^2 + 0.105\text{Cement} + 0.259\text{Water} - 0.001\text{Water}^2 - 0.008\text{Sand} - 0.025\text{Coarse} \qquad (5)$$

The error results from the calculations using the modeled equations are shown in Table 2. As the value of R2 of Multiple Linear Regression (MLR) is 0.8177 and the Multiple Non-Linear Regression value is 0.8791, which is close to 1, the MLNR is more accurate in terms of model development and the reliance on MLNR is preferred. In the case of MAE, the MLNR value is 4.642 and the MLR error value is 5.855; according to the criteria, lower values are preferred, which is why the MLNR is the leading model. The RMSE calculations show that the value of MLNR is 6.15, whereas the MLR error value is 7.43; in this error evaluation, the MLNR is more reliable because the RMSE also prefer values near 0.

Table 2. Error evaluation.

Model	R^2	MAE	RMSE
MLR	0.8177	5.855	7.43
MNLR	0.8791	4.642	6.15

4. Conclusions

The dataset is created using 115 results from the literature to train the model. The equations are developed separately for MLR and MNLR for the prediction of the compressive strength of NaOH-pre-treated crumb rubber concrete using different independent variables. The accuracy is analyzed using different errors like R^2, MAE, and RMSE. This study is conducted for the convenience of researchers to avoid laboratory procedures and extract direct results in a smart way.

The following conclusions can be drawn from this study:

- Several evaluation criteria are used for checking the accuracy of the models, including R^2, MAE, and RMSE. The MNLR performs the best compared to MLR;
- MLNR obtains the value of 0.8791, 4.642, and 6.15 for R^2, MAE, and RMSE, respectively;
- The determined sequence with respect to accuracy in this research is MNLR > MLR;
- Using a comprehensive set of variables for concrete mixture design, including alternative waste materials, was found to be feasible for predicting the strength of sustainable concrete.

The research needs some more studies, like parametric and sensitivity analyses, for more precise and practical implications in the real engineering world. After performing a detailed analysis using the parametric effect and sensitivity analysis; design engineers can use it with no objection.

Author Contributions: H.A.: Data collection, writing, modelling, formatting—original draft and validation. K.A.: Error calculations, writing, revised draft—review and editing. M.F.J.: resources, supervision and writing—revised draft. All authors have read and agreed to the published version of the manuscript.

Funding: This research received no external funding.

Institutional Review Board Statement: Not applicable.

Informed Consent Statement: Not applicable.

Data Availability Statement: Not applicable.

Conflicts of Interest: The authors declare no conflict of interest.

References

1. Ahmad, J.; Aslam, F.; Martinez-Garcia, R.; De-Prado-Gil, J.; Qaidi, S.M.A.; Brahmia, A. RETRACTED ARTICLE: Effects of waste glass and waste marble on mechanical and durability performance of concrete. *Sci. Rep.* **2021**, *11*, 21525. [CrossRef] [PubMed]
2. Kirthika, S.K.; Singh, S.; Chourasia, A. Alternative fine aggregates in production of sustainable concrete—A review. *J. Clean. Prod.* **2020**, *268*, 122089. [CrossRef]
3. Ren, F.; Mo, J.; Wang, Q.; Ho, J.C.M. Crumb rubber as partial replacement for fine aggregate in concrete: An overview. *Constr. Build. Mater.* **2022**, *343*, 128049. [CrossRef]
4. Qaidi, S.; Al-Kamaki, Y.S.S.; Al-Mahaidi, R.; Mohammed, A.S.; Ahmed, H.U.; Zaid, O.; Althoey, F.; Ahmad, J.; Isleem, H.F.; Bennetts, I. Investigation of the effectiveness of CFRP strengthening of concrete made with recycled waste PET fine plastic aggregate. *PLoS ONE* **2022**, *17*, 0269664. [CrossRef] [PubMed]
5. Tang, Q.; Ma, Z.; Wu, H.; Wang, W. The utilization of eco-friendly recycled powder from concrete and brick waste in new concrete: A critical review. *Cem. Concr. Compos.* **2020**, *114*, 103807. [CrossRef]
6. Ho, H.-J.; Iizuka, A.; Shibata, E. Chemical recycling and use of various types of concrete waste: A review. *J. Clean. Prod.* **2021**, *284*, 124785. [CrossRef]
7. Sankar, B.; Ramadoss, P. Modelling the compressive strength of high-performance concrete containing metakaolin using distinctive statistical techniques. *Results Control. Optim.* **2023**, *12*, 100241. [CrossRef]

8. Ahmed, H.U.; Abdalla, A.A.; Mohammed, A.S.; Mohammed, A.A.; Mosavi, A. Statistical Methods for Modeling the Compressive Strength of Geopolymer Mortar. *Materials* **2022**, *15*, 1868. [CrossRef] [PubMed]
9. Saloni; Parveen; Pham, T.M.; Lim, Y.Y.; Malekzadeh, M. Effect of pre-treatment methods of crumb rubber on strength, permeability and acid attack resistance of rubberised geopolymer concrete. *J. Build. Eng.* **2021**, *41*, 102448. [CrossRef]
10. Jokar, F.; Khorram, M.; Karimi, G.; Hataf, N. Experimental investigation of mechanical properties of crumbed rubber concrete containing natural zeolite. *Constr. Build. Mater.* **2019**, *208*, 651–658. [CrossRef]
11. Awan, H.H.; Javed, M.F.; Yousaf, A.; Aslam, F.; Alabduljabbar, H.; Mosavi, A. Experimental Evaluation of Untreated and Pretreated Crumb Rubber Used in Concrete. *Crystals* **2021**, *11*, 558. [CrossRef]
12. Mohammadi, I.; Khabbaz, H.; Vessalas, K. Enhancing mechanical performance of rubberised concrete pavements with sodium hydroxide treatment. *Mater. Struct.* **2016**, *49*, 813–827. [CrossRef]
13. Safan, M.; Eid, F.M.; Awad, M. Enhanced properties of crumb rubber and its application in rubberized concrete. *Int. J. Curr. Eng. Technol.* **2017**, *7*, 1784–1790.
14. Youssf, O.; Mills, J.E.; Hassanli, R. Assessment of the mechanical performance of crumb rubber concrete. *Constr. Build. Mater.* **2016**, *125*, 175–183. [CrossRef]
15. Youssf, O.; ElGawady, M.A.; Mills, J.E.; Ma, X. An experimental investigation of crumb rubber concrete confined by fibre reinforced polymer tubes. *Constr. Build. Mater.* **2014**, *53*, 522–532. [CrossRef]
16. Li, D.; Zhuge, Y.; Gravina, R.; Benn, T.; Mills, J.E. Creep and drying shrinkage behaviour of crumb rubber concrete (CRC). *Aust. J. Civ. Eng.* **2020**, *18*, 187–204. [CrossRef]
17. Youssf, O.; Mills, J.E.; Benn, T.; Zhuge, Y.; Ma, X.; Roychand, R.; Gravina, R. Development of Crumb Rubber Concrete for Practical Application in the Residential Construction Sector—Design and Processing. *Constr. Build. Mater.* **2020**, *260*, 119813. [CrossRef]

Disclaimer/Publisher's Note: The statements, opinions and data contained in all publications are solely those of the individual author(s) and contributor(s) and not of MDPI and/or the editor(s). MDPI and/or the editor(s) disclaim responsibility for any injury to people or property resulting from any ideas, methods, instructions or products referred to in the content.

Proceeding Paper

Comparative Seismic Response Analysis of a Multi-Storey Building with and without Base Isolators under High Magnitude Earthquake [†]

Maliha Mehar Qambrani *, Fizza Mirza and Muhammad Habib

Department of Civil Engineering, Balochistan University of Information Technology, Engineering and Management Sciences (BUITEMS), Quetta 87100, Pakistan; fizzamirza18@gmail.com (F.M.); habibcivilian05@gmail.com (M.H.)

* Correspondence: immqq16@gmail.com
† Presented at the 5th Conference on Sustainability in Civil Engineering (CSCE), Online, 3 August 2023.

Abstract: Earthquakes can induce structural failure, the vertical collapse of a structure, or can result in the breaking and falling of non-structural components of the structure. Most of the region of Pakistan has a high risk of seismic activity and the country lacks seismic-resistant structures. The effectiveness of the base isolation technique has been studied by many researchers, but significant research has not been conducted particularly for high earthquake-prone regions of Balochistan. This study includes the comparative seismic response analysis of the two multi-storey reinforced concrete 3D frames, with and without base isolation under the effect of seismic loads. This study indicates the effectiveness of base isolators in reducing the seismic response of buildings in the earthquake-prone regions of Balochistan.

Keywords: base isolation; FEM; seismic analysis; Uniform Building Code (UBC-97)

1. Introduction

An earthquake is a disruptive and unpredictable force that causes the Earth's surface to shake along a fault plane. They are the costliest disasters in history [1,2]. The 1935 Quetta earthquake destroyed the city completely, killing 30,000 to 60,000 in impact. Moreover, more than 450,000 buildings were damaged in the Kashmir Earthquake in 2005 [3]. Recently, two major quakes of magnitudes 7.5 and 7.8 occurred in south-central Turkey on 6 February 2023, near the Syrian border killing over 50,000 people [4]. The principal reason for structural failure during an earthquake is poor-quality construction materials, faults in construction methods, soil and foundation failure, mass irregularities, and inadequate design. With population growth, the demand for tall buildings is increasing. Worldwide, Reinforced Concrete (RC) multi-storied buildings have been affected by earthquakes drastically, and shear walls and steel bracing are used to reduce the earthquake effects but emerging technologies such as base isolators and dampers are found to be relatively more effective [5]. In earthquake-prone areas, there is always a risk of dangerous seismic activities. Several structures could collapse at magnitudes 6 to 7, with cracks on the ground. Many buildings may fall during earthquakes of magnitude 7 to 8 with significant damage [6]. Therefore, while designing a medium-to-high-rise building, vibration control should be considered along with the seismic design of the building. The effects of the horizontal component of an earthquake can be reduced by increasing the structure's natural period and decreasing the acceleration response which can be achieved utilizing seismic isolation [7]. Base isolators decouple the superstructure from the foundation, consisting of flexible or sliding materials placed between the building and its foundation. Extensive research shows that base isolation works properly for medium to high-rise buildings [8]. However, the novelty of this research is its focus on the response analysis of multi-storey buildings in regions

of Balochistan, with high risk of seismic activity. The significance of this work lies in its potential to contribute to future research and base isolator usage for earthquake protection in Balochistan.

2. Research Methodology

2.1. Modelling of Frame

Two five-storey identical building frames were modeled with and without base isolators shown in Figure 1a,b. The Lead Rubber Bearing (LRB) base isolator was modeled in ETABS by applying springs in the model at the location of joints. The link/support properties used in defining the springs are shown in Table 1 [9].

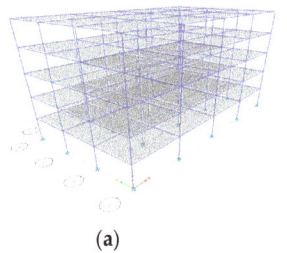
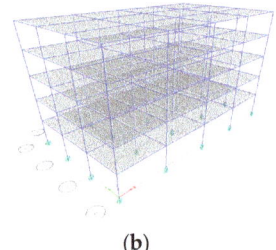

(a) (b)

Figure 1. 3D View of the SMRF frame: (**a**) without base isolation; (**b**) with base isolation.

Table 1. Lead rubber base isolator properties [9].

Lead Rubber Isolator Properties			
Rotational inertia	0.016603 kN/m	Distance from the end (Non-linear)	0.00318 m
Effective stiffness (Linear)	1,175,418.57 kN/m	Stiffness (Non-linear)	10,831 kN/m
Effective stiffness (Non-linear)	1175.42 kN/m	Yield strength (Non-linear)	34.70 kN
Effective Damping (Non-linear)	5%	Post-yield strength ratio	0.1

2.2. Ground Motion Data

A large magnitude earthquake, Tabas, Iran was used in seismic response analysis, obtained by selection of the best matched ground motion amongst different ground motions selected initially from PEER Ground Motion Database and through SeismoSignal. The properties of the ground motion used are shown in Table 2.

Table 2. Ground motion data [10].

Earthquake	Year	Magnitude	Fault Mechanism
Tabas, Iran	1978	7.35	Reverse

2.3. Ground Motion Parameters

Ground Motion parameters used for the seismic analysis and matching of the ground motion are selected based on UBC-97, which are according to the regions lying in the most critical zones of Balochistan, i.e., Zone 4. The parameters are shown in Table 3.

2.4. Matched Response Spectra and Time History

Initially, the UBC-97 design spectrum was generated in Seismomatch with a PGA value of 0.32 g, then matched to the UBC-97 design spectrum from which the response spectra and matched time history were obtained, as shown in Figures 2a and 2b, respectively.

Table 3. Ground motion parameters.

Soil Profile	SD	Stiff soil
Seismic zone factor	ZONE Z	4 0.4
Seismic Coefficient "Ca"	Soil Profile Type SD	Seismic zone factor, Z (Z = 0.4) 0.44Na
Seismic Coefficient "Cv"	Soil Profile Type SD	Seismic zone factor, Z (Z = 0.4) 0.64Na
Near Source factor "Na"	Seismic source type B Na	Closest distance to the known seismic source ≥10 km 1.0
Near Source factor "Nv"	Seismic source type B Na	Closest distance to the known seismic source ≥15 km 1.0

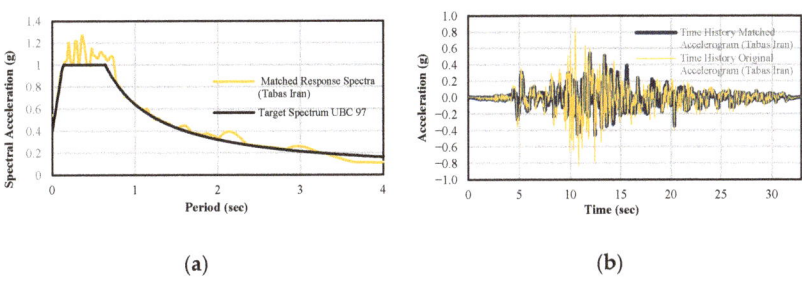

Figure 2. (**a**) Spectral acceleration (X-direction); (**b**) Tabas, Iran time history (X-direction).

3. Results and Discussions

3.1. Storey Displacement

The storey displacement is the absolute value of displacement of the storey with respect to the base. Figure 3a shows that the overall maximum inter-storey displacement has been significantly reduced. Hence, a reduction of 74.3% in roof storey displacement is observed.

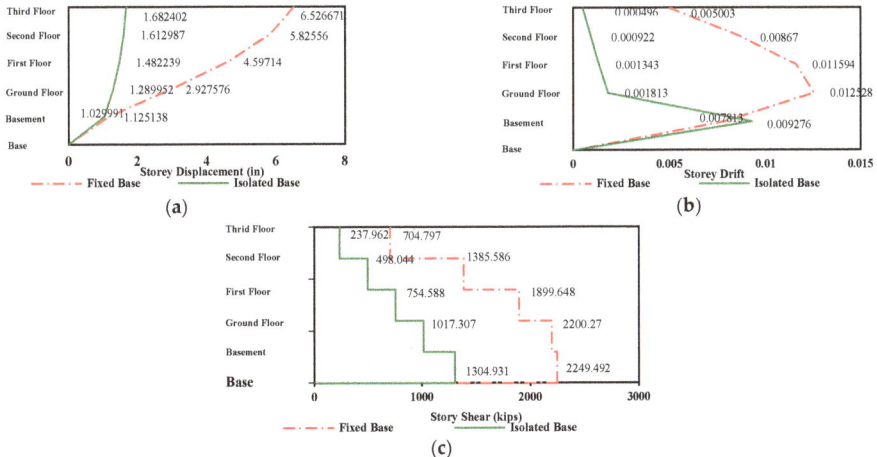

Figure 3. (X-direction) (**a**) Storey displacement; (**b**) Inter-storey drift; (**c**) Storey shear.

3.2. Inter-Storey Drift Ratio

Inter-storey drift ratio is the difference of displacements between two consecutive stories or storey drift, divided by the height of the story. The non-structural damage has a direct connection with the story drift [11]. At the roof storey, a reduction of 90.1% in the inter-storey drift ratio is observed. The following pattern can be seen in Figure 3b below.

3.3. Storey Shear

Storey shear is the lateral force caused by seismic and wind forces that are acting on a storey. Base isolation increases time period and storey drift while decreasing storey shear and acceleration, making stiffer structures more flexible by directing energy to the foundation [12]. Hence, a reduction of 66.2% in roof storey shear is observed in Figure 3c.

4. Conclusions

This study proves that base isolators significantly reduce storey displacements, drifts, and shear in multi-storey buildings in Balochistan's high-seismic-risk Zone 4. LRB isolators effectively increased stiffness, energy dissipation, and building time period from 0.8 s to 2.03 s, mitigating the earthquake impact by preventing resonance and reducing deformations.

Author Contributions: M.M.Q., F.M. and M.H. made equal and substantial contributions to all aspects encompassing the final manuscript. All authors have read and agreed to the published version of the manuscript.

Funding: This research received no external funding.

Institutional Review Board Statement: Not applicable.

Informed Consent Statement: Not applicable.

Data Availability Statement: No new data were created or analyzed in this study. Data sharing is not applicable to this article.

Conflicts of Interest: The authors declare no conflict of interest.

References

1. Kakpure, G.G.; Mundhada, A. Comparative study of static and dynamic seismic analysis of multistoried RCC buildings by ETAB. *Int. J. Eng. Res. Appl.* **2016**, *7*, 6–10.
2. Guin, J.; Saxena, V. *Extreme Losses from Natural Disasters-Earthquakes, Tropical Cyclones and Extratropical Cyclones*; Applied Insurance Research Inc.: Boston, MA, USA, 2000.
3. Farooqi, M.U.; Ali, M. Effect of pre-treatment and content of wheat straw on energy absorption capability of concrete. *Constr. Build. Mater.* **2019**, *224*, 572–583. [CrossRef]
4. Xinyu, J.; Xiaodong, S.; Tian, L.; Kaixin, W. Moment magnitudes of two large Turkish earthquakes on 6 February 2023 from long-period coda. *Earthq. Sci.* **2023**, *36*, 169–174.
5. Ahmadi, G. Overview of base isolation, passive and active vibraton control strategies for a seismic design of structure. *Sci. Iran.* **1995**, *2*, 18.
6. Asiala, D.W.P.C. UPSeis. Available online: https://www.mtu.edu/geo/community/seismology/learn/earthquake-measure/magnitude/ (accessed on 8 July 2023).
7. Torunbalci, N. Seismic isolation and energy dissipating systems in earthquake resistant design. In Proceedings of the 13th World Conference on Earthquake Engineering, Vancouver, BC, Canada, 1–6 August 2004.
8. Komuro, T.; Nishikawa, Y.; Kimura, Y.; Isshiki, Y. Development and Realization of Base Isolation System for High-Rise Buildings. *J. Adv. Concr. Technol.* **2005**, *3*, 233–239. [CrossRef]
9. Naeim, F.; Kelly, J.M. *Design of Seismic Isolated Structures: From Theory to Practice*; John Wiley & Sons: Hoboken, NJ, USA, 1999.
10. May, P.J.; Feeley, T.J.; Wood, R.; Burby, R.J. *Pacific Earthquake Engineering Research Center*; University of California: Berkeley, CA, USA, 2007.
11. Taghinezhad, R.; Taghinezhad, A.; Mahdavifar, V.; Soltangharaei, V. Evaluation of Story Drift under Pushover Analysis in Reinforced Concrete Moment Frames. *Int. J. Res. Eng.* **2018**, *5*, 296–302.
12. Kumar, G.J.S. Effect of Base Isolation using LRB on Stepped Building. *Ijraset J. Res. Appl. Sci. Eng. Technol.* **2013**, *10*, 4324–4330. [CrossRef]

Disclaimer/Publisher's Note: The statements, opinions and data contained in all publications are solely those of the individual author(s) and contributor(s) and not of MDPI and/or the editor(s). MDPI and/or the editor(s) disclaim responsibility for any injury to people or property resulting from any ideas, methods, instructions or products referred to in the content.

Proceeding Paper

Improvement of Early-Age Mechanical Properties of Cement Mortar by Adding Biochar of the Santa Maria Feverfew Plant [†]

Ahmed Kamal Subhani, Mohib Nisar and Anwar Khitab *

Department of Civil Engineering, Mirpur University of Science and Technology (MUST), Mirpur 10250, AJ&K, Pakistan; ahmedksubhani@gmail.com (A.K.S.); mohibsardar11@gmail.com (M.N.)
* Correspondence: anwar.ce@must.edu.pk
[†] Presented at the 5th Conference on Sustainability in Civil Engineering (CSCE), Online, 3 August 2023.

Abstract: This study focused on the application of nano-/micro-sized fibers obtained from pyrolysis of Santa Maria feverfew (biochar) in cement mortars. The biochar was added in amounts of 0, 0.05 and 0.1 percent by mass of cement. The mechanical characteristics were determined after 3 and 7 days and matched with those of the control samples. The compressive strength remained unchanged with the biochar addition, whereas the flexural strength increased. Biochar is a carbon-rich material, and its use in building materials leads to carbon sequestration, which is in accordance with the sustainable development goals of the UNO.

Keywords: cement mortars; biochar; Santa Maria feverfew plant; compressive strength; flexural strength; early age; carbon sequestration; sustainable development goals

1. Introduction

Pyrolysis is an energy-intensive method that involves the thermochemical breakdown of raw biomass in an inert atmosphere at high temperatures and pressures. This procedure yields different valuable products, like biochar, liquid bio-oil and fuel gases [1]. Biochar is a low-density dark-color carbon deposit. Scientists have used many different kinds of feedstock, including water hyacinth, oriental beech, corncob and many more [2–4]. Many researchers in the recent past have added biochar to cementitious products to enhance performance. Gupta et al. added biochar of sawdust at the rate 2% by mass of cement [5]. The results reveal that the addition enhanced the compressive strength and ductility of the end products. Tayyab et al. incorporated the biochar of millet and maize in mortar [6]. The authors reported an enhanced fracture toughness and ductility of the specimens, which was attributed to crack bridging/branching due to the fibrous nature of biochar. Iftekhar et al. studied the effect of adding the biochars of sugarcane bagasse and pine needles into cementitious mortars [7]. The authors reported enhanced interface shielding due to the addition. Restuccia et al. used the biochar of hazelnut shells as an additive in mortar specimens. They stated it enhanced compressive strength, flexural strength, toughness and ductility. Most of the previous studies focused on the effect of biochar on the hardened properties of the cementitious composites. The literature as regards the influence of biochar additions on the early-age properties of cementitious materials is limited.

The present study focused on the addition of the biochar of Santa Maria feverfew on the mechanical characteristics of cementitious mortar at an early age. As a matter of fact, the effect of biochar on the hardened properties of cementitious materials is well known; however, research on its effect on early-age mortar's characteristics is limited. Santa Maria feverfew is a local plant also known as gajjar boti or gajjar ghass. This study involved the evaluation of cementitious mortar in terms of its compressive and flexural strengths.

2. Materials and Methods

Cementitious mortars containing both the control mix and specimens with added biochar were prepared. A C-53-grade local ordinary Portland cement was used; the physical and chemical properties are described in Table 1. Sand was acquired from Lawrencepur (a well-known quarry) and was used as a fine aggregate, and its physical characteristics are reported in Table 1. Unlike the control mix, the other two mixes were integrated with biochar of the Santa Maria feverfew plant, as shown in Figure 1. Three types of specimens were prepared: one control, and those containing 0.05% and 0.1% biochar by mass of cement. The composition of the materials is presented in Table 2. A 1:1.5 cement/sand mortar with a water-to-cement ratio of 0.35 was prepared. The admixture was added at the rate of 1% by mass of cement.

Table 1. Chemical and physical characteristics of the ingredients of mortar.

Compound	%Age	Cement Property	Value	Sand Property	Value
CaO	61	Specific gravity	3.1	Specific gravity	2.7
SiO_2	21	Soundness	2%	Fineness modulus	2.7
Fe_2O_3	3	Fineness	1%	Bulk density (Kg/m^3)	1480
Al_2O_3	6	Initial setting time	30 min	Dry rodded bulk density (Kg/m^3)	1820
MgO	1.5	Final setting time	610 min	Water absorption(%)	3.9
Alkalis	0.5	Consistency	24%	Water content (%)	1.98
Gypsum	4				

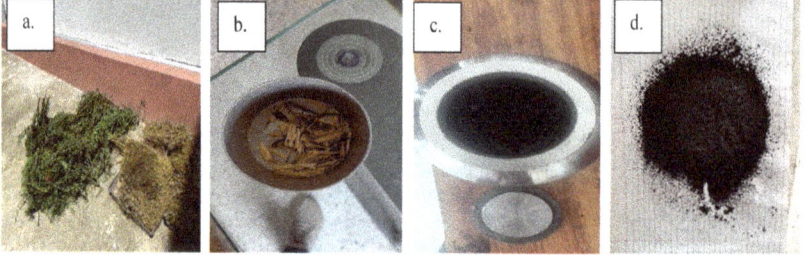

Figure 1. (a) Fresh Santa Maria feverfew plant, (b) dry plant, (c) biochar and (d) powdered biochar.

Table 2. Composition of mortar.

Samples	OPC (g)	Sand (g)	Water (mL)	Biochar (g)	Admixture (mL)	W/C
C0	610	916	214	0	6	0.35
C0.05	610	916	214	0.305	6	0.35
C0.1	610	916	214	0.61	6	0.35

All the materials as described in Table 2 were mixed as per the ASTM C305-20 standard method [8]. To avoid agglomeration caused by the fine size of the biochar particles, the biochar was mixed in water using the UV-sonication technique. The admixture (super-plasticizer) was added (1% by mass of cement) to make the dispersion more effective. All the ingredients were mixed in a Hobart mixer. After mixing, the samples were cast in cubes (50 mm size) and prisms (40 × 40 × 160 mm) according to the ASTM C1314 method [9]. After a one-day period, the specimens were de-molded and immersed in water for curing. The compressive strength was measured through ASTM C109 [10]. The flexural strength of the specimens was determined using the ASTM C348 method [11]. The mixing machine, molding process, strength test and sonicated mix of biochar and water are shown in Figure 2.

Figure 2. (a) Mixing, (b) molding, (c) flexural strength test and (d) UV-sonicated solution.

3. Results and Discussion

3.1. Compressive Strength

The strength results are presented in Table 3. The results show that there was a slight decrease in the compressive strength. The compressive strength of cementitious composites is closely related to their density [12]. Being porous and lightweight, biochar particles reduce density. A reduction in density might lead to a reduction in compressive strength [13,14]. In the hardened state, biochars are observed to enhance compressive strength [15,16].

Table 3. Effect of biochar on compressive strength of cementitious mortar.

Specimen	Compressive Strength (MPa)		% Difference
	3 Days	7 Days	
C0	13.2	16.7	
C0.05	12.3	15.3	−8
C0.1	12.5	15.8	−5

3.2. Flexural Strength

The influence of the biochar on the flexural strength of the cementitious mortar is shown in Table 4. The flexural strength was enhanced with the biochar addition. A previous study suggests that while the compressive strength mainly relies on the compactness of the material, the flexural strength is mainly dependent on the bond between the ingredients of cementitious materials [17]. As micro-/nano-fibers like those composing biochar enhance the cohesion between the particles, they may enhance the flexural strength [18].

Table 4. Effect of biochar on flexural strength of cementitious mortar.

Specimen	Flexural Strength (MPa)		% Difference
	3 Days	7 Days	
C0	1.3	1.9	
C0.05	1.7	2.4	26
C0.1	2.1	2.7	42

4. Conclusions

Based on the experimental outputs, the following conclusions are put forward. The biochar of the Santa Maria feverfew plant slightly reduces the compressive strength at an early age. The reduction in material density seems to be the cause of the reduction in compressive strength. The compressive strength is reduced by 8 and 5% with 0.05 and 0.1% additions of the biochar. Biochar enhances the flexural strength at an early age. Enhancement of material cohesion due to biochar's fibrous character seems to be the cause of the enhancement of flexural strength. The flexural strength is enhanced by 26 and 42% with 0.05 and 0.1% additions of the biochar at 7 days. Biochars are highly carbon-rich particles. As such, their addition to cementitious mixes is beneficial for carbon capture, a key sustainable development goal. Its addition results in high flexural strength of the end products at an early age, which is beneficial in many construction projects.

Author Contributions: Conceptualization, A.K.; methodology, A.K.S. and M.N.; validation, A.K.S. and M.N.; formal analysis, A.K.S. and M.N.; investigation, A.K.S. and M.N.; resources, A.K.; data curation, A.K.S.; writing—original draft preparation, A.K.; writing—review and editing, A.K.; visualization, M.N.; supervision, A.K.; project administration, A.K.; funding acquisition, A.K. All authors have read and agreed to the published version of the manuscript.

Funding: This research received external funding from Higher Education Commission of Pakistan through NRPU-7984 project.

Institutional Review Board Statement: Not applicable.

Informed Consent Statement: Not applicable.

Data Availability Statement: The authors confirm that the data supporting the findings of this study are available within the article.

Conflicts of Interest: The authors declare no conflict of interest.

References

1. Khitab, A.; Ahmad, S.; Khan, R.A.; Arshad, M.T.; Anwar, W.; Tariq, J.; Khan, A.S.R.; Khan, R.B.N.; Jalil, A.; Tariq, Z. Production of Biochar and Its Potential Application in Cementitious Composites. *Crystals* **2021**, *11*, 527. [CrossRef]
2. Carnaje, N.P.; Talagon, R.B.; Peralta, J.P.; Shah, K.; Paz-Ferreiro, J. Development and characterisation of charcoal briquettes from water hyacinth (Eichhornia crassipes)-molasses blend. *PLoS ONE* **2018**, *13*, e0207135. [CrossRef]
3. Aburas, H.; Demirbas, A. Evaluation of beech for production of bio-char, bio-oil and gaseous materials. *Process Saf. Environ. Prot.* **2015**, *94*, 29–36. [CrossRef]
4. Gupta, G.K.; Ram, M.; Bala, R.; Kapur, M.; Mondal, M.K. Pyrolysis of chemically treated corncob for biochar production and its application in Cr(VI) removal. *Environ. Prog. Sustain. Energy* **2018**, *37*, 1606–1617. [CrossRef]
5. Gupta, S.; Kua, H.W.; Low, C.Y. Use of biochar as carbon sequestering additive in cement mortar. *Cem. Concr. Compos.* **2018**, *87*, 110–129.
6. Tayyab, S.; Khitab, A.; Iftikhar, A.; Khan, R.B.N.; Kirgiz, M.S. Manufacturing of high-performance light-weight mortar through addition of biochars of millet and maize. *Waste Dispos. Sustain. Energy* **2023**, *5*, 97–111.
7. Iftikhar, A.; Arsalan Khushnood, R.; Khitab, A.; Ahmad, S. Improved fracture response and electromagnetic interference shielding effectiveness of cementitious composites incorporating pyrolytic bagasse fibers and pine needles. *Constr. Build. Mater.* **2023**, *370*, 130722.
8. *ASTM C305*; Standard Practice for Mechanical Mixing of Hydraulic Cement Pastes and Mortars of Plastic Consistency. ASTM International: West Conshohocken, PA, USA, 2022.
9. *ASTM C1314-21*; Standard Test Method for Compressive Strength of Masonry Prisms. ASTM International: West Conshohocken, PA, USA, 2021.
10. *ASTM C109/C109M-16a*; Standard Test Method for Compressive Strength of Hydraulic Cement Mortars (Using 2-in. or [50-mm] Cube Specimens). ASTM International: West Conshohocken, PA, USA, 2016. [CrossRef]
11. *ASTM ASTM C348*; Standard Test Method for Flexural Strength of Hydraulic-Cement Mortars1. ASTM International: West Conshohocken, PA, USA, 1998.
12. Arif, R.; Khitab, A.; Kırgız, M.S.; Khan, R.B.N.; Tayyab, S.; Khan, R.A.; Anwar, W.; Arshad, M.T. Experimental analysis on partial replacement of cement with brick powder in concrete. *Case Stud. Constr. Mater.* **2021**, *15*, e00749. [CrossRef]
13. Fityus, S.; Imre, E. The significance of relative density for particle damage in loaded and sheared gravels. *EPJ Web Conf.* **2017**, *140*, 07011. [CrossRef]
14. Setina, J.; Gabrene, A.; Juhnevica, I. Effect of Pozzolanic Additives on Structure and Chemical Durability of Concrete. *Procedia Eng.* **2013**, *57*, 1005–1012.
15. Khushnood, R.A.; Ahmad, S.; Restuccia, L.; Spoto, C.; Jagdale, P.; Tulliani, J.M.; Ferro, G.A. Carbonized nano/microparticles for enhanced mechanical properties and electromagnetic interference shielding of cementitious materials. *Front. Struct. Civ. Eng.* **2016**, *10*, 209–213. [CrossRef]
16. Khushnood, R.A.; Ahmad, S.; Ferro, G.A.; Restuccia, L.; Tulliani, J.M.; Jagdale, P. Modified fracture properties of cement composites with nano/micro carbonized bagasse fibers. *Frat. Integrità Strutt.* **2015**, *34*, 534–542.

17. Holm, T.A.; Bremner, T.W. *State-of-the-Art Report on High-Strength, High-Durability Structural Low-Density Concrete for Applications in Severe Marine Environments*; ERDC/SL TR-00-3; U.S. Army Engineer Research and Development Center: Vicksburg, MI, USA, 2000.
18. Khitab, A.; Ahmad, S.; Khushnood, R.A.; Rizwan, S.A.; Ferro, G.A.; Restuccia, L.; Ali, M.; Mehmood, I. Fracture toughness and failure mechanism of high performance concrete incorporating carbon nanotubes. *Frat. Integrità Strutt.* **2017**, *11*, 238–248. [CrossRef]

Disclaimer/Publisher's Note: The statements, opinions and data contained in all publications are solely those of the individual author(s) and contributor(s) and not of MDPI and/or the editor(s). MDPI and/or the editor(s) disclaim responsibility for any injury to people or property resulting from any ideas, methods, instructions or products referred to in the content.

Proceeding Paper

Evaluation of Seismic Response of 3D Building Frame with and without Base Isolation Using Finite Element Analysis [†]

Fahad Ali, Hammad Azam * and Muhammad Habib

Department of Civil Engineering, Balochistan University of Information Technology, Engineering and Management Sciences (BUITEMS), Quetta 87100, Pakistan; faahadali39@gmail.com (F.A.); habibcivilian05@gmail.com (M.H.)
* Correspondence: hammadazam13@gmail.com; Tel.: +92-3357551324
† Presented at the 5th Conference on Sustainability in Civil Engineering (CSCE), Online, 3 August 2023.

Abstract: An earthquake is a force which is unpredictable and can cause serious damage to a structure and its structural components. For improved safety, the seismic design of the building should be adequate, and measures should be taken for vibration control. In order to mitigate effects, base-isolation techniques must be provided to counter seismic loads; this is a seismic isolation technique which prevents the transfer of energy from the base of the structure to the upper stories. In the research, two similar 3D frame models were modelled with and without base isolators and analyzed following the provisions and codes. The real ground motion data of the earthquake was selected, then matched to a design spectrum to obtain the matched time history. The story response plots obtained after the time history analysis indicate that base isolation is a reliable and effective technology to improve the seismic performance of the building, particularly with inadequate seismic design.

Keywords: base isolators; seismic base isolation; lead rubber base isolator; time history analysis

1. Introduction

The health of structures during seismic activity in seismic-prone areas has always been a concern for humanity. An earthquake, which mostly occurs at fault zones, is the shaking of the ground caused by the seismic waves that travel through the Earth's crust. Seismic waves are produced when the energy stored in the Earth's crust is released suddenly, usually when rock masses strain against each other fractures and slip. Nowadays, in the modern construction world, seismic isolation techniques have gained a lot of popularity. This technique has been adopted to protect the structures against the adverse effects of seismic activity [1]. By placing isolation devices between the foundation and super-structure, it effectively decouples the structure from seismic ground vibrations. The base isolation device's primary goal is to reduce the horizontal acceleration which is transmitted to the structure [2]. Earthquakes of lower magnitudes, while perceived as less severe, can still pose significant dangers. The most effective way to protect buildings from seismic forces is to make the structure flexible and increase its stiffness, which is achieved by base isolators.

Base isolation systems are generally of two types, i.e., elastomeric bearings and sliding isolation bearings. The elastomeric bearing is a type in which synthetic rubber is sandwiched between two mild steel plates as a damping object, while a sliding isolation bearing is a flexible device that allows some movement by the column end which slides over a plate, it deflects the earthquake energy [3]. Studies have shown that the horizontal component of an earthquake plays a major role in the destruction of a structure; it is more responsible for the damage to the structures. It can be reduced by increasing the structure's natural period and decreasing the acceleration response, which can be achieved using base isolators [4]. In 2017, Wankhade studied the behavior of different base isolators in a building and compared peak story shear and displacements with and without base isolators. He concluded that there was a considerable difference in peak story displacements and shear

with fixed and isolated bases and that base frequency was reduced by using base isolators. The base isolation system is not physically present in Pakistan, but rather, studies have been conducted on the feasibility and application of lead rubber bearings and different kinds of isolations. There are many earthquake parameters used to determine the ground motion characteristics. Generally, earthquake loading effects are represented by three parameters, that include: peak ground acceleration, time history, and response spectrum. These parameters are used for the design of the structure and are known as the Design Basis Ground Motion Parameters [5].

2. Methodology

The experiment starts with modelling two similar 3D frame models in CSI ETABS version 2016. The elevation, plan, and 3D view of both the frames are shown in Figure 1a–d. Both frames shared the same design parameters and characteristics. Isolation was incorporated at the base of the second frame. The input parameters were complied with the UBC-97 specifications and ACI-318 requirements. The models were analyzed based on a finite element analysis. Moreover, for base isolators, lead rubber bearings were used because they are commonly used in buildings and are easy to install, compared to others. In ETABS base isolator acted as spring element. The input parameters for isolators are given in Table 1.

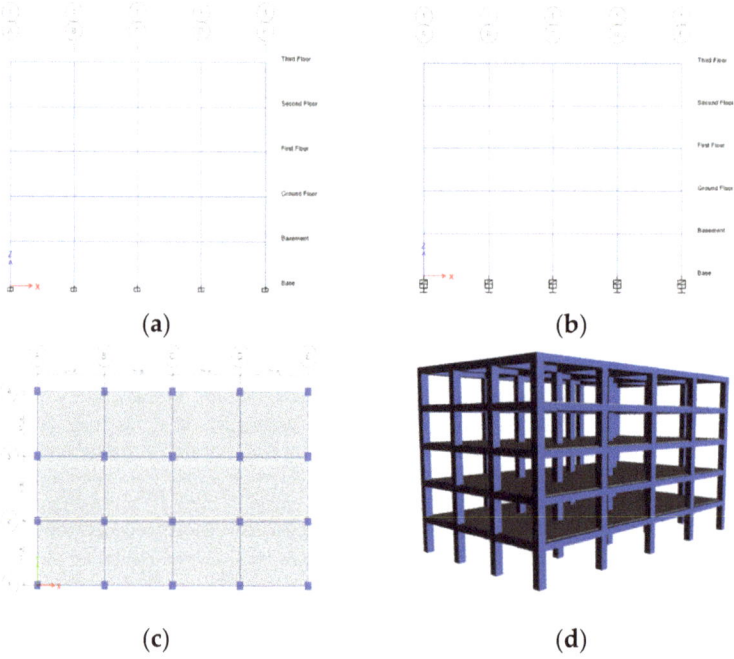

Figure 1. This figure shows the elevations, plan and 3D views of the frame models: (**a**) shows the elevation of frame without base isolation, while (**b**) shows with base isolation; (**c**) shows plan view of the frame model while (**d**) shows 3D view of the frame model.

Table 1. Stiffness characteristics for the design of lead rubber isolator [6].

Lead Rubber Isolator (LRB) Properties	
Rotational inertia	0.009721 kN/m
Effective stiffness (Linear)	889,950.58 kN/m
Effective stiffness (Non-linear)	889.95 kN/m
Effective damping (Non-linear)	5%
Stiffness (Non-linear)	8201 kN/m
Yield strength (Non-linear)	26.27 kN
Post yield strength ratio	0.1

3. Ground Motion Time History

The earthquake used for analysis was the Northridge—Lassen and Reseda earthquake that occurred in 2007, with a magnitude of 4.66 and having a reverse fault mechanism. The earthquake was matched to the UBC-97 design spectrum using Seismomatch, which was then used in time history analysis to obtain the story response plots. The time history can be seen below in Figure 2.

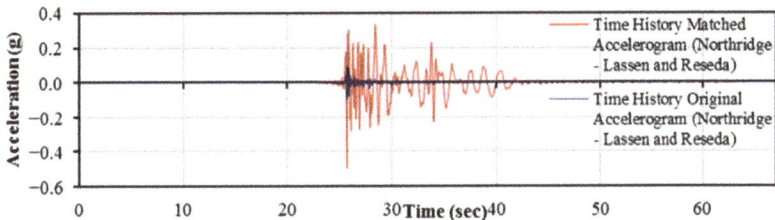

Figure 2. Northridge—Lassen and Reseda time history (acceleration in X-direction).

4. Seismic Response

The comparison of the story displacement of both the fixed base and the isolated base is shown in Figure 3. The base isolation systems cause the superstructure to move rigidly, which reduces the relative structural element displacement and, in effect, reduces internal forces on beams and columns. For the earthquake, i.e., the Northridge—Lassen and Reseda, a significant reduction of 85.9% in roof story displacement was observed, as shown in Figure 1a. When comparing the base-isolated frame to the fixed base frame, it can be seen in Figure 3b that the story drift ratio is higher on lower floors, and then it becomes constant on upper floors. However, in the fixed base frame, the story drift starts at zero at the ground level and increases nonlinearly with height. A reduction of 94.5% in roof story drift has been observed. Story shear refers to the lateral or horizontal force exerted on each level or story of a building during an earthquake event. Compared to the fixed base frame, a significant reduction in shear at each story is observed, and a reduction in roof story shear of 80.7% was observed in the base-isolated frame, as shown in Figure 3c.

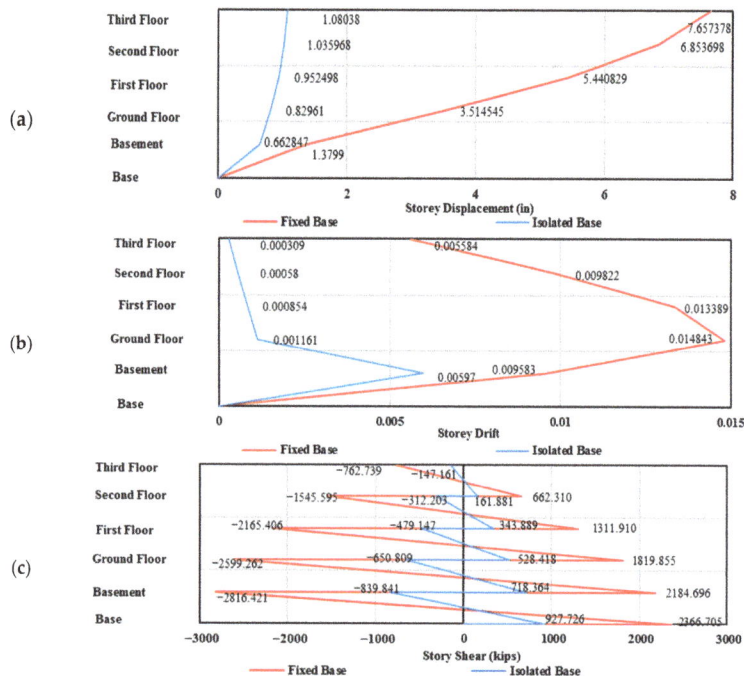

Figure 3. This figure shows the seismic response of the building frame in terms of story response plots: (**a**) shows the story displacement for Northridge—Lassen and Reseda for fixed and isolated base models (X-direction), (**b**) shows the story drift, while (**c**) shows the story shear.

5. Conclusions

The conclusions drawn from the results of seismic response analysis are as follows:

- In comparison to a fixed-base frame, the frame with base isolation shows a significant decrease in story displacement, i.e., by 85.9%; a story drift of 94.5%; and story shear, which is 80.7% because of the reduction in seismic effect after the provision of lead rubber base isolators.
- The results indicate that for earthquakes of relatively lower magnitudes, base isolators have been proven to be very effective and important, especially for buildings which have not been designed according to the seismic demand.
- This study demonstrates that base isolation can be successfully used in mid-rise buildings to mitigate the effects of earthquakes and can be used for retrofitting to prevent structural damage.

Author Contributions: Conceptualization, methodology, software, and validation, F.A. and H.A.; investigation, F.A.; data curation, F.A.; writing—original draft preparation, H.A.; writing—review and editing, H.A.; visualization and supervision, M.H. All authors have read and agreed to the published version of the manuscript.

Funding: This research received no external funding.

Institutional Review Board Statement: Not applicable.

Informed Consent Statement: Not applicable.

Data Availability Statement: Not applicable.

Conflicts of Interest: The authors declare no conflict of interest.

References

1. Liu, X.; Chen, J.; Wang, H.; Jia, Z.; Wu, Z. Earthquake Economic Loss Assessment of Reinforced Concrete Structures Using Multiple Response Variables. *Buildings* **2023**, *13*, 1719. [CrossRef]
2. Patil, S.J.; Reddy, G.R. State of Art Review-Base Isolation Systems for Structures. *Int. J. Emerg. Technol. Adv. Eng.* **2012**, *2*, 438–453.
3. Ryan, K.L.; Kelly, J.M.; Chopra, A.K. Nonlinear model for lead–rubber bearings including axial-load effects. *J. Eng. Mech.* **2005**, *131*, 1270–1278. [CrossRef]
4. Torunbalci, N. Seismic isolation and energy dissipating systems in earthquake resistant design. In Proceedings of the 13th World Conference on Earthquake Engineering, Vancouver, BC, Canada, 1–6 August 2004; Citeseer: Forest Grove, OR, USA, 2014.
5. Muruva, H.P.; Kiran, A.; Bandyopadhyay, S.; Reddy, G.; Agrawal, M.; Verma, A.K. Design Basis Ground Motion. In *Textbook of Seismic Design*; Springer: Berlin/Heidelberg, Germany, 2019; pp. 29–60.
6. Naeim, F.; Kelly, J.M. *Design of Seismic Isolated Structures: From Theory to Practice*; John Wiley & Sons: New York, NY, USA, 1999.

Disclaimer/Publisher's Note: The statements, opinions and data contained in all publications are solely those of the individual author(s) and contributor(s) and not of MDPI and/or the editor(s). MDPI and/or the editor(s) disclaim responsibility for any injury to people or property resulting from any ideas, methods, instructions or products referred to in the content.

Proceeding Paper

Validation of Chlorine Decay Equation for Water Quality Analysis in Distribution Networks [†]

Rehan Jamil [1], Hamidi Abdul Aziz [2,*] and Mohamad Fared Murshed [1]

1. School of Civil Engineering, Engineering Campus, Universiti Sains Malaysia, Gelugor 14300, Pulau Pinang, Malaysia; rehan.jamil@student.usm.my (R.J.); cefaredmurshed@usm.my (M.F.M.)
2. Solid Waste Management Cluster, Universiti Sains Malaysia, Gelugor 14300, Pulau Pinang, Malaysia
* Correspondence: cehamidi@usm.my
† Presented at the 5th Conference on Sustainability in Civil Engineering (CSCE), Online, 3 August 2023.

Abstract: This article discusses the extent of the spread of contamination in water distribution networks which may enter through a pipe leak, and the decay rate of chlorine for a specified design duration. A comprehensive water quality analysis is performed using EPANet 2.2 for the spread of contamination and chlorine decay. The results show that a contaminant entering at the highest point of the network would pollute the whole network whereas the effect of such a contaminant would be limited if it enters at the lowest location. Also, the initial chlorine concentration is found to be more for such critical nodes which are higher in elevation, although the decay rate remains the same. The research proves to be beneficial for the management of water distribution through pipe networks against contaminants for maintaining public health.

Keywords: contamination; chlorine decay; water distribution networks; decay rate

1. Introduction

A leak in the water supply pipe may cause contamination to enter into the network which presents serious harm to living beings [1–3]. Pipe leakage may occur due to corrosion of pipe material, incorrect laying of pipes, or any impact or excessive loads on the pipes [4,5]. The contamination which enters into the pipe may be of two types namely, reactive or non-reactive. The reactive matter includes toxic substances or pathogen bacteria or related soil particles which stay inside the pipes and make the water harmful for the users [1,6]. To resolve this, disinfectants are added to the water distribution networks. Chlorine (Cl) being the most common disinfectant is used worldwide for the said purpose. However, it entails various limitations, as the Cl reacts with the water in the pipe, called bulk reaction, or reacts with the pipe wall material, called the wall reaction. Due to these reactions, the concentration of Cl in water reduces with time and this phenomenon is termed Cl decay [7]. The concentration of Cl is designed in such a way that it follows the guideline provided by the World Health Organization. LeChevalliar et al. studied the health risks related to contamination intrusion in water networks [2]. Keramat et al. performed numerical analysis for the contamination intrusion due to pressure transients [8]. Darweesh found that there is a negative effect on the water quality when using variable speed pumps for the water supply network [9]. Ponti et al. and Li et al. used evolutionary algorithms and a multi-parameter monitoring system to detect intrusion and its location in distribution networks, respectively [10,11]. Currently, there is a need for validation of the equation of chorine decay rate and the amount of contamination that may spread into the network at a given time duration after it enters the network. Hence, this study was performed to study; the extent of contaminant spread into the water network and the decay rate of contaminant while being in the water network for a designed duration.

2. Research Methodology

2.1. Study Area

The satellite imagery of the study area is shown in Figure 1a. The valley is a residential area with an approximate population of 3150 and it consists of houses of various sizes including other amenity buildings. The elevation contour plan of the area is shown in Figure 1b. The network model and its all design parameters have been obtained from the website of Aquaveo [12]. The primary source of water for these valleys is the Mann Water Reservoir located on the south-eastern side of the area [12]. The circular water tank (having a diameter of 115 ft (35 m) with a water depth of 10 ft (3 m) is designed to be filled by pumps and the location of the water tank is kept at an elevated area with an elevation of 6320 ft (1925 m) such that the pressure at all joints in the network remains sufficient. The water distribution network (WDN) laid for the area consists of High-Density Poly Ethylene (HDPE) pipe material with a pressure rating of PN-16 (maximum sustainable pressure of 16 bars or 230 psi). The network model in the EPANet 2.2 consists of a total of 116 nodes and 119 pipes. The water distribution pipes range from 3 in. to 6 in. (75 mm to 160 mm) in diameter and the layout used for the simulation is shown in Figure 2.

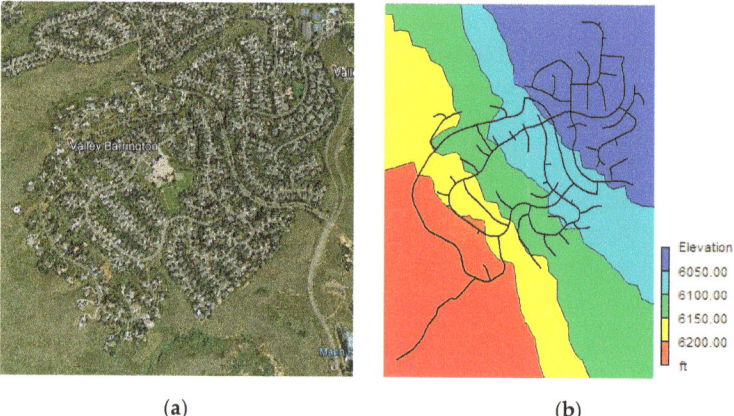

(a)   (b)

Figure 1. (a) Satellite image and (b) elevation contour map of the study area.

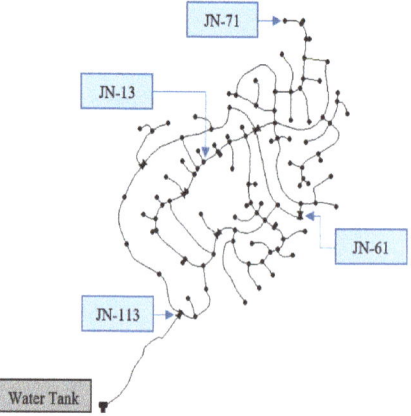

Figure 2. Water supply network for Denver with critical nodes.

2.2. Hydraulic Analysis Parameters

Figure 2 shows the WDN with the location of identified critical nodes that were identified based on their water demand and elevation values. JN-13 was selected for having the highest demand of 7.2 gpm in the network whereas JN-61 was selected for having the lowest demand and no outflow. Also, JN-71 was selected for being the lowest junction in the whole network system with an elevation of 5986 ft (1824 m). The steady-state hydraulic analysis was run by using the Hazen-Williams equation. A roughness coefficient of 150 was used for plastic pipes [13]. The following equation was used for the analysis: $h_L = \frac{4.73 L Q^{1.85}}{C^{1.85} D^{4.87}}$, where h_L represents the head loss in the pipe, L is the length of the pipe in ft, Q is the volume of water in cfs, C is the Hazen-Williams roughness coefficient, and D represents the diameter of the pipe in ft. The decay of any disinfectant, such as Cl, in WDN, is the function of first-order kinetics as shown in Equation by [14]: $C_t = C_i e^{-kt}$, where C_t is the concentration of Cl after a specific duration t, C_i is the initial concentration of Cl, and k is the coefficient of decay.

3. Results and Discussion

3.1. Contaminant Spread

Contour plots of residual Cl were prepared to see the spread of contamination through each node at the end of the design duration and well-justified results were obtained, as shown in Figure 3. The plot obtained from JN-13 having the highest demand shows a very limited spread of the organic material present in the contaminant, which can be compared with the plot obtained for JN-61 having the lowest demand and shows a huge spread of the organic matter. As of JN-61, the contaminant is seen to have traveled even upstream into the network, making it the critical location in the pipe network. Moreover, JN-71, being at the lowest elevation level, shows almost no spread and can be compared to JN-113, being at the highest point in the network and showing the full spread of organic material into the network. If the contaminant enters into a network from the highest point in elevation in the network, the organic matter is likely to spread in the whole network making it the most critical point. The results obtained by the contamination spread plots are justified and validated by obtaining the bulk reaction reports from the EPANet 2.2 program shown in Table 1. The spread percentage of 79.61 shown by the bulk reaction reports indicates that when the contaminant enters at the extreme upstream of the WDN at JN-113, it misbalances almost the whole network and all the downstream pipe junctions are affected.

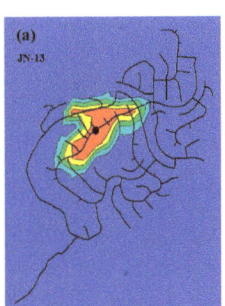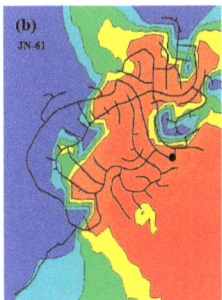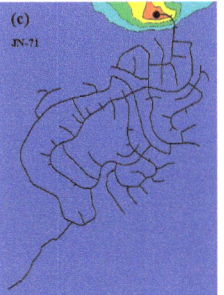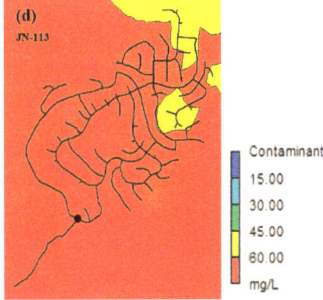

Figure 3. Contour plots of contaminant spread at critical nodes shown as dot.

Table 1. Contaminant spread percentage in the network considered entry at each node.

Critical Pipe Junctions	Contamination Spread (%)
JN-13	5.01
JN-61	21.58
JN-71	3.07
JN-113	79.61

3.2. Chlorine Decay Rate

The analysis of Cl decay was performed using the trial and error method. The assumed initial Cl concentration was set in the system and the levels of Cl were monitored for the whole design duration of 168 h. When the level of Cl was found to be above the WHO recommendation at the end of the analysis, the initial concentration was reduced, and the analysis was run again till the values fell within the range. By using the values obtained through simulation, the equation of the trend line for contamination decay at JN-113 was found to be as shown in Figure 4 with an r^2 value of 1, $C_t = \frac{11.7}{e^{0.016t}}$.

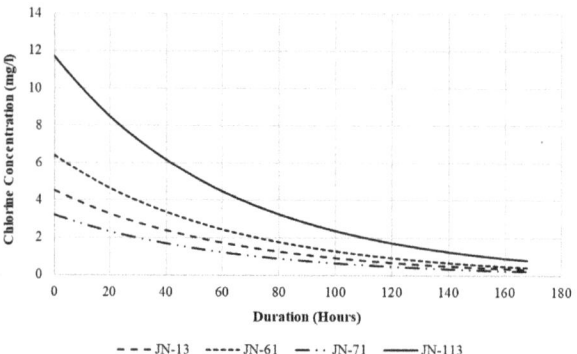

Figure 4. Chlorine decay pattern at each node.

The initial Cl concentration at JN-113 was found to be 11.7 mg/L. The coefficient of determination (r^2) value for the trend line shows that the data perfectly fit the equation of the line. Similarly, the initial Cl concentrations for JN-13, JN-61, and JN-71 were found to be 4.5 mg/L, 6.4 mg/L, and 3.2 mg/L, respectively. By studying the plots, shown in Figure 4, it was found that the Cl decay rate follows an inverse exponential path for all nodes although the values differ from one node to another.

4. Conclusions

Hydraulic analysis for contamination spread and chlorine decay was performed for a WDN by injecting the contaminant at four identified critical nodes separately. The results of contaminant intrusion analysis present that a contaminant entering at a higher level would spread up to the locations in the network which are lower than the source of contamination. On the other hand, if the contaminant enters the lowest point in the network, it will remain limited to that point and not likely to contaminate the whole system. Also, it was observed that contaminant spread was more at the nodes having lower base demands and such nodes showed more spread of the contaminant. Moreover, it was found that the critical points in the network, responsible for higher contamination spread, need higher initial chlorine concentrations. The chlorine decay rates were found to be the same for all the critical points and the decay rate follows an inverse exponential relation relating to the time. The critical points in a water supply network mentioned in this research can be monitored regularly for any damage or irregularity to avoid any incident of contaminant intrusion. Also, the quality of chlorine to be added to the water can be regularized by using the results of this research. This research can be very useful for the management, planning, and maintenance of water distribution networks.

Author Contributions: Conceptualization, R.J. and H.A.A.; methodology, R.J. and M.F.M.; software, R.J.; validation, H.A.A. and M.F.M.; resources, H.A.A. and M.F.M.; data curation, R.J.; writing—original draft preparation, R.J.; writing—review and editing, H.A.A. and M.F.M.; supervision, H.A.A. and M.F.M.; project administration, H.A.A. All authors have read and agreed to the published version of the manuscript.

Funding: This research received no external funding.

Institutional Review Board Statement: Not applicable.

Informed Consent Statement: Not applicable.

Data Availability Statement: The datasets generated and analyzed in the research are available from the corresponding author and can be furnished upon request.

Conflicts of Interest: The authors declare no conflict of interest.

References

1. Boyd, G.R.; Wang, H.; Britton, M.D.; Howie, D.C.; Wood, D.J.; Funk, J.E.; Friedman, M.J. Intrusion within a Simulated Water Distribution System due to Hydraulic Transients I: Volumetric Method and Comparison of Results. *J. Environ. Eng.* **2004**, *130*, 774–777. [CrossRef]
2. LeChevallier, M.W.; Gullick, R.W.; Karim, M.R.; Friedman, M.; Funk, J.E. The Potential for Health Risks from Intrusion of Contaminants into the Distribution System from Pressure Transients. *J. Water Health* **2003**, *1*, 3–14. [CrossRef] [PubMed]
3. Gong, J.; Guo, X.; Yan, X.; Hu, C. Review of Urban Drinking Water Contamination Source Identification Methods. *Energies* **2023**, *16*, 705. [CrossRef]
4. Qasem, A.; Jamil, R. GIS-Based Financial Analysis Model for Integrated Maintenance and Rehabilitation of Underground Pipe Networks. *J. Perform. Constr. Facil.* **2021**, *35*, 1–8. [CrossRef]
5. Rodríguez, J.M.; Galván, X.D.; Ramos, H.M.; Amparo, P.A.L. An overview of leaks and intrusion for different pipe materials and failures. *Urban Water J.* **2014**, *11*, 1–10. [CrossRef]
6. Karim, M.R.; Abbaszadegan, M.; Lechevallier, M. Potential for Pathogen Intrusion during Pressure Transients. *J. Am. Water Work. Assoc.* **2003**, *95*, 134–146. [CrossRef]
7. Tamminen, S.; Ramos, H.; Covas, D. Water Supply System Performance for Different Pipe Materials Part I: Water Quality Analysis. *Water Resour. Manag.* **2008**, *22*, 1579–1607. [CrossRef]
8. Keramat, A.; Payesteh, M.; Brunone, B.; Meniconi, S. Interdependence of flow and pipe characteristics in transient induced contamination intrusion: Numerical Analysis. *J. Hydroinform.* **2020**, *22*, 473–490. [CrossRef]
9. Darweesh, M. Impact of variable speed pumps on water quality in distribution systems. *Water SA* **2018**, *44*, 419–427. [CrossRef]
10. Ponti, A.; Candelieri, A.; Giordani, I.; Archetti, F. Intrusion Detection in Networks by Wasserstein Enabled Many-Objective Evolutionary Algorithms. *Mathematics* **2023**, *11*, 2342. [CrossRef]
11. Li, Z.; Liu, H.; Zhang, C.; Fu, G. Generative adversarial networks for detecting contamination events in water distribution systems using multi-parameter, multi-site water quality monitoring. *Environ. Sci. Ecotechnol.* **2023**, *14*, 100231. [CrossRef] [PubMed]
12. Aquaveo, L.L.C. 2023. Available online: https://www.aquaveo.com/ (accessed on 13 April 2023).
13. Jamil, R. Frictional Head Loss Relation between Hazen-Williams and Darcy-Weisbach Equations for Various Water Supply Pipe Materials. *Int. J. Water* **2019**, *13*, 333–347. [CrossRef]
14. Jamil, R. Performance of a New Pipe Material UHMWPE against Disinfectant Decay in Water Distribution Networks. *Clean Technol. Environ. Policy* **2018**, *20*, 1287–1296. [CrossRef]

Disclaimer/Publisher's Note: The statements, opinions and data contained in all publications are solely those of the individual author(s) and contributor(s) and not of MDPI and/or the editor(s). MDPI and/or the editor(s) disclaim responsibility for any injury to people or property resulting from any ideas, methods, instructions or products referred to in the content.

Proceeding Paper

The Behavior of Pre-Treated Crumb Rubber and Polypropylene-Fiber-Incorporated Mortar Subjected to Elevated Temperatures [†]

Manail Shafqat [1,*], Muhammad Basit Khan [1] and Hamad Hassan Awan [2]

[1] Department of Civil Engineering, COMSATS University Islamabad, Abbottabad Campus, Abbottabad 22060, Pakistan; niazisq67@gmail.com
[2] School of Engineering and Digital Sciences, Nazarbayev University, Astana 010000, Kazakhstan; hamad.awan@nu.edu.kz
* Correspondence: manahilshafqat123@gmail.com
[†] Presented at the 5th Conference on Sustainability in Civil Engineering (CSCE), Online, 3 August 2023.

Abstract: Rubber is a waste product produced by the industrial sector in large quantities. Due to its non-degradable nature, it has been a serious threat to the environment. Thus, it is recommended to develop concrete or mortar containing rubber, so that it can save our environment, and it is economical too. Crumb rubber, when incorporated in mortar, reduces its strength, so it can be used along with some fibers to enhance its strength. This study examined the effect of elevated temperatures, i.e., 150, 300, 450, 600, and 750 °C, on mortar samples containing 5% crumb rubber replacement of fine aggregate by volume, and with the incorporation of 1% PPF. The findings indicated a rise in compressive strength up to 300 °C, followed by a subsequent decline. It was also observed that the weight loss of the samples increased with an increase in temperature.

Keywords: muffle furnace; polypropylene fibers (PP fibers); compressive strength; elevated temperature

1. Introduction

Advancements in infrastructure have led to increased waste production from demolished structures and increasing population, including plastic and rubber tires [1]. These materials persist in the environment, causing environmental issues [2]. Tire reclamation is a major problem, with stockpiled scrap rubber tires posing fire hazards and potential breeding grounds for pests [3]. Alternative uses include fueling cement production, producing carbon-black in asphalt, or aggregates in cementitious composites. Replacing natural aggregates in cementitious composites with waste rubber tires could prevent environmental pollution and make it economically feasible [1]. However, CR's hydrophobic nature and surface sleekness cause weak matrix bonding [4–6]. Pre-treatment chemicals like lime, NaOH, and detergents can resolve this issue [7].

Researchers have studied cementitious materials incorporating recycled rubber tires (RRA) at elevated temperatures, finding reduced compressive and tensile strength. However, rubberized concrete remained strong at 200 °C and 400 °C [8–10]. Polypropylene fibers improve cement matrix mechanical traits after fire and exposure to extreme temperatures. They enhance the flexural strength and crack resistance but have a lower flexural strength and crack resistance than regular mortar [11–13].

There are no studies currently available that incorporate both crumb rubber and polypropylene fibers subjected to elevated temperatures. In this study, the behavior of pre-treated crumb rubber and PP-fiber-incorporated mortar subjected to elevated temperatures is assessed. Various proportions of crumb rubber have been employed as a replacement for fine aggregate and PP fibers are introduced in addition to the mortar. The main objective of this study is to use different pre-treatment methods and to find the optimum percentages.

The optimum PPFs are also checked. The collective effect of both optimal conditions is examined on the mechanical properties of the mortar specimens.

2. Research Methodology

Specimens underwent a 28-day curing period at room temperature and were tested for resistance to elevated temperatures (150 °C, 300 °C, 450 °C, 600 °C, and 750 °C) as shown in Figure 1. Samples were placed in a muffle furnace after 28 days of curing. The temperature increased at a rate of 5 °C/min and remained at the desired temperature for 30 min. The furnace then cooled down naturally.

(a) (b) (c)

Figure 1. (a) Samples placed in a muffle furnace, (b) samples after heating at excessive temperature, and (c) samples after being taken out from the muffle furnace.

Mix Design

This study employed a mix design of 1:2.75 (cement to aggregate ratio) as per ASTM C39 shown in Table 1. Cubes with dimensions of 50 mm × 50 mm × 50 mm were prepared for the samples. To enhance compressive strength, 5% lime-treated crumb rubber replacement and 1% PP fibers were added. Crumb rubber replacement was performed by volume, and the optimum concentration of treated lime with PP fibers is referred to as PPFL.

Table 1. Mix design for the PPFL samples.

Samples	Mix Ratio	Cement (g)	Sand (g)	W/C ratio	Water (g)	CR (mL)	PPF (g)
PPFL (Compression)	1:2.75	500	1306.25	0.48	240	50	7.158

3. Experimental Procedures

3.1. Treatment of CR with Lime, NaOH, and Water

Crumb rubber underwent pre-treatments to enhance its mechanical properties. For lime treatment, a 10% lime solution was used, and the crumb rubber was immersed in it for 24 h. It was then washed and sun-dried. NaOH treatment involved dipping the crumb rubber in a 10% NaOH solution for 24 h, followed by washing and sun-drying. Water treatment included boiling the crumb rubber for 24 h, after which it was dried.

3.2. Polypropylene Fibers (PP Fibers)

PP fibers were added to balance out the reduction in mechanical properties, and the inclusion of PP fibers increased the strength.

3.3. Compressive Strength and Weight Loss

To calculate the weight loss, the initial weight of the samples was recorded after a 28-day curing period. The final weight was measured after subjecting the samples to elevated temperatures and collecting them from the oven. The formula for weight loss is

$$Weight\ loss = \frac{initial\ wt. - final\ wt.}{initial\ wt.} \times 100 \qquad (1)$$

4. Results

Weight Loss and Compressive Strength

Table 2 illustrates that weight loss increases with higher muffle furnace temperatures, indicating sample crack development and disintegration. The highest weight loss of 8.97% is observed at 750 °C, while the lowest is 1.087% at 150 °C. Initially, the compressive strength is lower at 150 °C, increases at 300 °C due to melted crumb rubber releasing hydrostatic pressure, and then decreases again. This study's maximum compressive strength is 25.597 MPa at 300 °C, while the minimum is 5.64 MPa at 750 °C.

Table 2. Weight loss and compressive strength of mortar samples at elevated temperature.

Sample No	Muffle Furnace Temp	Weight before Elevated Temp	Weight after Elevated Temp	Compression Test	Weight Loss (%)
1	150 °C	269.93 g	266.99 g	20.56 MPa	1.08%
2	150 °C	271.69 g	268.69 g	19.78 MPa	1.10%
3	150 °C	277.15 g	274.14 g	21.08 MPa	1.08%
1	300 °C	279.58 g	264.64 g	24 MPa	5.34%
2	300 °C	276.78 g	262.45 g	25.87 MPa	5.18%
3	300 °C	274.03 g	259.54 g	26.92 MPa	5.28%
1	450 °C	269.2 g	252.69 g	16.52 MPa	6.13%
2	450 °C	277 g	258.95 g	17.56 MPa	6.51%
3	450 °C	279 g	261.16 g	18.36 MPa	6.39%
1	600 °C	276 g	252.18 g	13.16 MPa	8.63%
2	600 °C	279.76 g	255.19 g	13.32 MPa	8.78%
3	600 °C	278.95 g	254.46 g	12.28 MPa	8.77%
1	750 °C	275.93 g	259.81 g	5.6 MPa	7.84%
2	750 °C	276.68 g	250.28 g	4.12 MPa	9.54%
3	750 °C	277.62 g	251.17 g	7.2 MPa	9.53%

Figure 2a indicates the behavior of compressive strength over temperature. From 150 °C to 300 °C, the compressive strength increases, but after increasing the temperature further, it decreases gradually. This shows that the compressive strength is ideal at 300 °C. Figure 2b shows that in the initial stage of increasing temperature, there is a sharp increase in weight loss, but further temperature increases have a small impact on weight loss. By increasing the temperature from 150 °C to 300 °C, the weight loss rises to 5%, but a further increase of 150 °C causes a 1–2% weight loss.

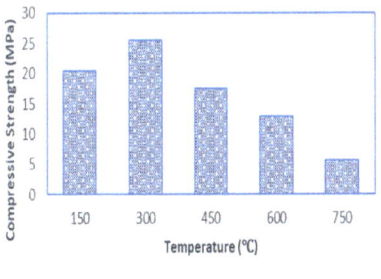
(a)

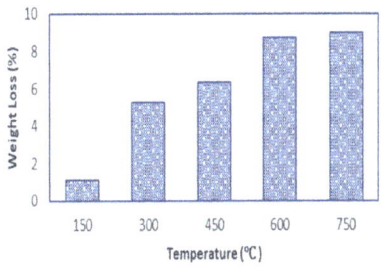
(b)

Figure 2. (a) Compressive strength values at elevated temperature. (b) Weight loss at elevated temperature.

5. Conclusions

This study concludes that

- Among the four treatments, lime treatment was found to be most effective, while water treatment yielded the worst results.
- The optimum percentage of crumb rubber replacement was found to be 5%, while the optimal PPF addition percentage was 1%.
- Rubberized-polypropylene-fiber-reinforced mortar can be exposed to temperatures up to 300 °C because, after this temperature, the compressive strength started decreasing.
- Initially, the strength increased from 150 to 300 °C. This may be due to the melting of CR, which may act as a paste. Also, with increasing temperature, hydrostatic pressure is generated, which acts against the load.
- The higher the temperature to which the sample is exposed, the greater the weight loss.
- The substitution of sand with CR had a significant impact on compressive strength, with a decrease observed as the proportion of CR increased. The results obtained in this study conclude that rubberized-polypropylene-fiber-reinforced mortar can be used in false facades, interior construction, road barriers, sideways, crash barriers around bridges, etc.

6. Recommendations

This study is inadequate to comprehend the actions of rubber-based cement composites and fiber-infused cementitious composites at high temperatures in a systematic manner. Further investigation is required to examine the mechanism and impact of the two additional fibers in the process.

Author Contributions: M.S.: literature review, writing, methodology, formatting—original draft, validation. M.B.K.: experimental work, writing, revised draft—review and editing. H.H.A.: resources, supervision, writing—revised draft. All authors have read and agreed to the published version of the manuscript.

Funding: This research received no external funding.

Institutional Review Board Statement: Not Applicable.

Informed Consent Statement: Not Applicable.

Data Availability Statement: Not Applicable.

Conflicts of Interest: The authors declare no conflict of interest.

References

1. Ahmad, J.; Zhou, Z.; Majdi, A.; Alqurashi, M.; Deifalla, A.F. Overview of Concrete Performance Made with Waste Rubber Tires: A Step toward Sustainable Concrete. *Materials* **2022**, *15*, 5518. [CrossRef] [PubMed]
2. Batayneh, M.; Marie, I.; Asi, I. Use of selected waste materials in concrete mixes. *Waste Manag.* **2007**, *27*, 1870–1876. [CrossRef] [PubMed]
3. Siddique, R.; Naik, T.R. Properties of concrete containing scrap-tire rubber—An overview. *Waste Manag.* **2004**, *24*, 563–569. [CrossRef] [PubMed]
4. Bravo, M.; de Brito, J. Concrete made with used tyre aggregate: Durability-related performance. *J. Clean. Prod.* **2012**, *25*, 42–50. [CrossRef]
5. Huynh, H.; Raghavan, D. Durability of simulated shredded rubber tire in highly alkaline environments. *Adv. Cem. Based Mater.* **1997**, *6*, 138–143. [CrossRef]
6. Fedroff, D.; Ahmad, S.; Savas, B. Mechanical Properties of Concrete with Ground Waste Tire Rubber. *Transp. Res. Rec.* **1996**, *1532*, 66–72. [CrossRef]
7. Awan, H.H.; Javed, M.F.; Yousaf, A.; Aslam, F.; Alabduljabbar, H.; Mosavi, A. Experimental Evaluation of Untreated and Pretreated Crumb Rubber Used in Concrete. *Crystals* **2021**, *11*, 558. [CrossRef]
8. Correia, J.R.; Marques, A.M.; Pereira, C.M.C.; de Brito, J. Fire reaction properties of concrete made with recycled rubber aggregate. *Fire Mater.* **2012**, *36*, 139–152. [CrossRef]
9. Guo, Y.-C.; Zhang, J.-H.; Chen, G.-M.; Xie, Z.-H. Compressive behaviour of concrete structures incorporating recycled concrete aggregates, rubber crumb and reinforced with steel fibre, subjected to elevated temperatures. *J. Clean. Prod.* **2014**, *72*, 193–203. [CrossRef]

10. Guelmine, L.; Hadjab, H.; Benazzouk, A. Effect of elevated temperatures on physical and mechanical properties of recycled rubber mortar. *Constr. Build. Mater.* **2016**, *126*, 77–85. [CrossRef]
11. Signorini, C.; Sola, A.; Malchiodi, B.; Nobili, A.; Gatto, A. Failure mechanism of silica coated polypropylene fibres for Fibre Reinforced Concrete (FRC). *Constr. Build. Mater.* **2020**, *236*, 117549. [CrossRef]
12. Choumanidis, D.; Badogiannis, E.; Nomikos, P.; Sofianos, A. The effect of different fibres on the flexural behaviour of concrete exposed to normal and elevated temperatures. *Constr. Build. Mater.* **2016**, *129*, 266–277. [CrossRef]
13. Zhang, Y.; Zhang, S.; Jiang, X.; Zhao, W.; Wang, Y.; Zhu, P.; Yan, Z.; Zhu, H. Uniaxial tensile properties of multi-scale fiber reinforced rubberized concrete after exposure to elevated temperatures. *J. Clean. Prod.* **2023**, *389*, 136068. [CrossRef]

Disclaimer/Publisher's Note: The statements, opinions and data contained in all publications are solely those of the individual author(s) and contributor(s) and not of MDPI and/or the editor(s). MDPI and/or the editor(s) disclaim responsibility for any injury to people or property resulting from any ideas, methods, instructions or products referred to in the content.

Proceeding Paper

Efficiency and Sustainability: Enhancing Mortar Mixtures with Wastepaper Sludge Ash †

Asad Shafique *, Ahsin Ihsan and Muhammad Faisal Javed

Department of Civil Engineering, Comsats University Islamabad, Abbottabad Campus, Abbottabad 22020, Pakistan; 500ahsin@gmail.com (A.I.); arbabfaisal@cuiatd.edu.pk (M.F.J.)
* Correspondence: 2asad.shafique@gmail.com
† Presented at the 5th Conference on Sustainability in Civil Engineering (CSCE), Online, 3 August 2023.

Abstract: The study aims to increase the efficiency of mortar mixes and improve their necessary qualities such as strength, density, and durability by using wastepaper as a cement substitute in the form of wastepaper sludge ash (WPSA). Mortars with 20, 25, and 30% cement replacement were tested. Due to less use of cement and greater usage of WPSA, CO_2 and SO_2 emissions can be reduced. The chemical properties of WPSA were compared to those of Ordinary Portland Cement (OPC). Testing showed that WPSA had similar cementitious properties. Results demonstrate the potential applications of this mortar in a variety of settings where increased toughness and equivalent characteristics are needed while still preserving the environment.

Keywords: wastepaper sludge ash; mortar; density; water absorption; acid attack

1. Introduction

The global population rapidly increases with time, which puts increased pressure on urban construction including residential, commercial, and industrial buildings. This has led to an increase in the need for cement use worldwide [1,2]. One of the most versatile fundamental construction materials is cement [3,4]. However, the cement sector is characterized by significant levels of energy consumption [5,6] and greenhouse gas emissions [7,8]. Around the globe, the cement industry emits roughly 7% of carbon dioxide (CO_2) [8,9]. Therefore, in order to lessen this influence on the environment, it is necessary to investigate a viable cement substitute [10]. Reusing waste materials rather than disposing of them in landfills is a practical and affordable solution to these problems [11,12]. Paper recycling industries produce wastes called wastepaper sludge and wastepaper sludge ash (WPSA). The chemical composition of WPSA varies; it usually includes lime (CaO), silica (SiO_2), and alumina (Al_2O_3) and can be used as a supplemental cementitious material (SCM) [13]. The primary objective of this research is to develop a green technology material. WPSA has been used in concrete, bricks, and in studies instead of mortar. Our goal is to develop an eco-friendly and effective mortar for construction purposes. Mortar is also used to repair joints and cracks. By replacing 25% of cement with WPSA in mortar mixes we can obtain such efficient mortar.

2. Materials and Methods

2.1. Materials

A proportional mixture of Ordinary Portland cement (OPC) and wastepaper sludge ash (WPSA) was used as a cementitious material. Wastepaper sludge was collected from private schools and universities of Abbottabad. It was dried in the sun for 12 to 15 days and then burnt in an electric furnace at 750 °C for 2 h to make ash. The composition of the WPSA, analyzed via X-Ray Fluorescence (XRF), is presented in Table 1. It consists primarily of calcium and can serve as a binding material.

Table 1. Chemical/elemental composition of WPSA.

Empirical Formula	WPSA (%)
Ca	75.37
Si	17.096
S	4.397
Fe	1.994
K	0.428
Ti	0.263
Sr	0.184
Zn	0.114
Cu	0.072
Mn	0.049
Zr	0.027
V	0.007

2.2. Methods

Mortar mix ratios for 'MC' samples for each ratio were prepared according to ASTM-C109 by partially replacing 0, 20, 25, and 30% of cement with WPSA by weight. The sand-to-binder ratio was taken as 1:2.75 and the water-to-cement ratio used for mixing was restricted to 0.48. Weights and proportioning data are shown in Table 2.

Table 2. Mix proportion of WPSA, cement, sand, and water.

Name	Cement	Sand	WPSA	Water
MC-0	500 g	1375 g	0 g	242 g
MC-20	400 g	1375 g	100 g	242 g
MC-25	375 g	1375 g	125 g	242 g
MC-30	350 g	1375 g	150 g	242 g

3. Results

3.1. Flowability

The study conducted showed that when the percentage of WPSA increased, the workability of the mortar decreased. Figure 1 displays the results of this research.

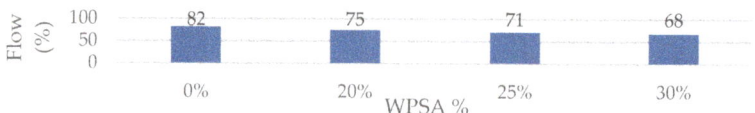

Figure 1. Flow/workability of 0, 20, 25, 30% WPSA mortar mix.

3.2. Compressive Strength

Mortar cubes with 25% WPSA exhibited favourable compressive strengths of 19.1 (CTM) and 18 MPa (RH). In comparison, the cubes with 20% and 30% WPSA replacements showed lower compressive strengths as shown in Figure 2. Therefore, replacing 25% of cement with WPSA results in a mixture that can be used for environmentally friendly construction purposes.

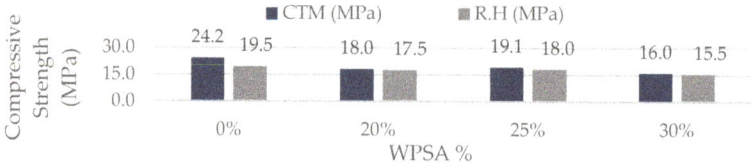

Figure 2. Compressive strength results (CTM and RH).

3.3. Ultrasonic Pulse Velocity

The test was conducted according to ASTM C597-02. The UPV results for WPSA replaced samples are shown in Figure 3. These results suggest that the 25% WPSA sample had higher velocity, indicating the most uniform mass, and similar quality.

Figure 3. Ultrasonic pulse velocity results for 150 mm long samples.

3.4. Water Absorption

The test was conducted according to ASTM C642-9. It was found that the addition of 25% WPSA to mortar mixtures resulted a significantly lower water absorption rate compared to other replacements, as depicted in Figure 4.

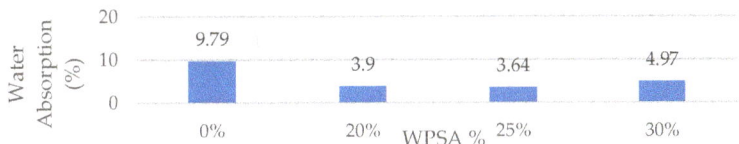

Figure 4. Water absorption results for 0, 20, 25, 30% WPSA mortar samples.

When 25% of the cement was replaced with WPSA, it increased the volume of the mixture and reduced internal voids, resulting in a denser mixture as given in Table 3.

Table 3. Bulk density values for WPSA mortar samples.

WPSA (%)	Density (g/cm^3)
0%	2.174
20%	2.18
25%	2.204
30%	2.065

3.5. Acid Attack (HCL)

In this study, HCL with a pH value of 3.01 was used to test the durability of the mortar samples. The results of the acid durability test showed that the acid resistance of the WPSA samples slightly decreased as the WPSA content increased. The results are shown in Figure 5.

Figure 5. Loss of strength (%) in WPSA mortar samples.

4. Conclusions

1. The chemical properties of WPSA showed similar behaviour to that of cement. A high calcium content in the WPSA results in good quality and characteristics of the mortar.
2. Replacement of 25% WPSA with cement showed favourable ultimate compressive strength and reduced water absorption by approximately 60%.

3. The technique developed in this study can be used in runoff structures, sewage pipes, canal surfaces, and wall plastering.

Author Contributions: Conceptualization, A.S. and M.F.J.; methodology, A.S.; validation, A.S., A.I. and M.F.J.; formal analysis, A.S.; investigation M.F.J.; resources, A.I.; data curation, A.S.; writing—original draft preparation, A.S.; writing—review and editing, A.S.; supervision, M.F.J. All authors have read and agreed to the published version of the manuscript.

Funding: This research received no external funding.

Institutional Review Board Statement: Not applicable.

Informed Consent Statement: Not applicable.

Data Availability Statement: Not applicable.

Conflicts of Interest: The authors declare no conflict of interest.

References

1. Alves, A.V.; Vieira, T.F.; De Brito, J.; Correia, J.R. Mechanical properties of structural concrete with fine recycled ceramic aggregates. *Constr. Build. Mater.* **2014**, *64*, 103–113. [CrossRef]
2. Nasr, M.S.; Hussain, T.; Kubba, H.; Shubbar, A.A. Influence of Using High Volume Fraction of Silica Fume on Mechanical and Durability Properties of Cement Mortar. *J. Eng. Sci. Technol.* **2020**, *15*, 2494–2506.
3. Azevedo, A.R.G.; Cecchin, D.; Carmo, D.F.; Silva, F.C.; Campos, C.M.O.; Shtrucka, T.G.; Marvila, M.T.; Monteiro, S.N. Analysis of the compactness and properties of the hardened state of mortars with recycling of construction and demolition waste (CDW). *J. Mater. Res. Technol.* **2020**, *9*, 5942–5952. [CrossRef]
4. Marvila, M.T.; Azevedo, A.R.G.; Monteiro, S.N. Verification of the application potential of the mathematical models of lyse, abrams and molinari in mortars based on cement and lime. *J. Mater. Res. Technol.* **2020**, *9*, 7327–7334. [CrossRef]
5. Atmaca, A.; Yumrutaş, R. Analysis of the parameters affecting energy consumption of a rotary kiln in cement industry. *Appl. Therm. Eng.* **2014**, *66*, 435–444. [CrossRef]
6. Amaral, L.F.; Girondi Delaqua, G.C.; Nicolite, M.; Marvila, M.T.; de Azevedo, A.R.G.; Alexandre, J.; Fontes Vieira, C.M.; Monteiro, S.N. Eco-friendly mortars with addition of ornamental stone waste—A mathematical model approach for granulometric optimization. *J. Clean. Prod.* **2020**, *248*, 119283. [CrossRef]
7. Pacheco-Torgal, F.; Jalali, S. Compressive strength and durability properties of ceramic wastes based concrete. *Mater. Struct. Constr.* **2011**, *44*, 155–167. [CrossRef]
8. De Azevedo, A.R.G.; Alexandre, J.; Xavier, G.d.C.; Pedroti, L.G. Recycling paper industry effluent sludge for use in mortars: A sustainability perspective. *J. Clean. Prod.* **2018**, *192*, 335–346. [CrossRef]
9. Abed, M.; Nasr, M.; Hasan, Z. Effect of silica fume/binder ratio on compressive strength development of reactive powder concrete under two curing systems. *MATEC Web Conf.* **2018**, *162*, 10–13. [CrossRef]
10. Shubbar, A.A.; Al-Jumeily, D.; Aljaaf, A.J.; Alyafei, M.; Sadique, M.; Mustafina, J. Investigating the mechanical and durability performance of cement mortar incorporated modified fly ash and ground granulated blast furnace slag as cement replacement materials. In Proceedings of the 12th International Conference on Developments in eSystems Engineering (DeSE), Kazan, Russia, 7–10 October 2019; pp. 434–439. [CrossRef]
11. Jang, H.S.; Lim, Y.T.; Kang, J.H.; So, S.Y.; So, H.S. Influence of calcination and cooling conditions on pozzolanic reactivity of paper mill sludge. *Constr. Build. Mater.* **2018**, *166*, 257–270. [CrossRef]

12. De Azevedo, A.R.G.; Alexandre, J.; Pessanha, L.S.P.; Manhães, R.d.S.T.; de Brito, J.; Marvila, M.T. Characterizing the paper industry sludge for environmentally-safe disposal. *Waste Manag.* **2019**, *95*, 43–52. [CrossRef] [PubMed]
13. García, R.; Vigil de la Villa, R.; Vegas, I.; Frías, M.; Sánchez de Rojas, M.I. The pozzolanic properties of paper sludge waste. *Constr. Build. Mater.* **2008**, *22*, 1484–1490. [CrossRef]

Disclaimer/Publisher's Note: The statements, opinions and data contained in all publications are solely those of the individual author(s) and contributor(s) and not of MDPI and/or the editor(s). MDPI and/or the editor(s) disclaim responsibility for any injury to people or property resulting from any ideas, methods, instructions or products referred to in the content.

Proceeding Paper

Soil Improvement Using Waste Polyethylene Terephthalate (PET) †

Tariq Saeed * and Muhammad Usman Arshid

Department of Civil Engineering, University of Engineering and Technology (UET), Taxila 47040, Pakistan; usman.arshid@uettaxila.edu.pk
* Correspondence: engrtariq88@gmail.com
† Presented at the 5th Conference on Sustainability in Civil Engineering (CSCE), Online, 3 August 2023.

Abstract: This study investigates the use of waste polyethylene terephthalate (PET) bottle strips for soil stabilization in the Potohar region. Uncontrolled filling during housing societies development has led to settlement issues and structural cracking. By incorporating PET bottle strips in varying compositions, the engineering properties of the soil were improved, including increased maximum dry density, bearing capacity, and unconfined compression strength. This research paper offers an innovative technique to mitigate settlement problems and presents an eco-friendly waste management solution for sustainable development.

Keywords: bearing capacity; maximum dry density; PET bottle strips; plate load test

1. Introduction

Recently soil stabization using polymers or waste materials such as polythene bags and waste plastic bottles are being explored by various researchers. A recent stud assesses the potential of cement kiln dust and plastic strips to enhance the properties of clayey soil. The inclusion of CKD increases the maximum dry density (MDD) of dune sand. An increase of 34% was achieved by mixing CKD [1,2]. An evaluation of waste marble dust has been carried out and reported to produce a considerable improvement in the physical properties of soil [3]. The plastic strips were of varying lengths (1 cm, 2 cm, and 3 cm) and different proportions of 0.2%, 0.5%, and 0.8%; an optimal improvement in the dry weight of soil was achieved with 2 cm plastic strips at 0.8% of the dry weight of the soil [4]. The fiber-reinforced soil improved the strength and engineering properties of the soil; the best percentage the plastic fiber achieved was 0–5%, there was an increased CBR value, and a reduction in the settlement of the dimensions, showing a higher aspect ratio to obtain better results [5]. The plastic strips were cut into different sizes, with lengths ranging from 12 mm to 21 mm and widths of 3 mm and 6 mm. Different concentration of PET content (0%, 0.4%, 0.6%, 0.8%, and 1% by soil weight) were incorporated. The highest unconfined compressive strength (UCS) was the one containing 0.8% PET strips with a width of 3 mm and a length of 18 mm; they achieved an optimum UCS 2.17 time compared to raw soil [6]. The liquefaction susceptibility of PET fiber-reinforced fine sand has been presented on the basis of results obtained through a series of cyclic triaxial tests. The number of cycles was four to reach liquefaction compared to unreinforced sand at a composition of 0.6% PET plastic fiber [7].

2. Materials and Methods Section

2.1. Sieve Analysis and Atterberg's Limit Test

The sieve analysis test was performed in accordance with ASTM D-6913. A 250 g soil sample was taken and washed through a sieve with a mesh size of 0.075 mm. After drying the sample for 24 h, the sample retained on sieve no. 200 was passed through a series of sieves in descending order, ranging from 4.75 mm to 0.075 mm. The Atterberg's limits tests were performed in accordance with ASTM D-4318. The liquid limit and plastic limit tests were performed on raw soil and composite samples with plastic strips contents of 0.4%, 0.7%, and 1.0%.

2.2. Standard Proctor and Plate Load Test

The standard proctor tests were performed following ASTM D-698. The soil sample was compacted in a 4″ diameter mold in three equal layers. Each layer was compacted by applying 25 blows with a hammer of 5.5 lb weight. The plate load tests were performed as per ASTM D-1194. A pit was excavated with dimensions of 2.50′ × 2.50′ × 2.0′ feet. The tests were performed on loosely filled soil, partially compacted soil, well-compacted pure soils, and well-compacted composite soils with plastic strips content of 0.4%, 0.7%, and 1.0%.

3. Research Methodology

Plastic bottles were collected from the disposal point. The head and tail of the bottles were cut, and a tool was used to convert the bottles into spirals. These spirals were then further cut into final strips of different sizes. The PET bottles were cut into strips of 3 × 6 mm, 3 × 9 mm, and 3 × 18 mm composition at 0.4%, 0.7%, and 1.0%. The stepwise procedure for cutting the waste plastic bottles into strips is presented in Figure 1.

Figure 1. Waste PET bottle strip process: (**a**) strip cutting tool; (**b**) PET spiral; (**c**) plastic strips sizes.

4. Results and Discussion

4.1. Sieve Analysis Test and Index Properties

The soil sample was collected locally followed by evaluation of the index and physical properties of the raw soil as shown in Table 1. The soil was classified according to the Unified Soil Classification System (USCS) as CL-ML, belonging to the Silty Clay group. The gradation curve of the back fill material is shown in Figure 2.

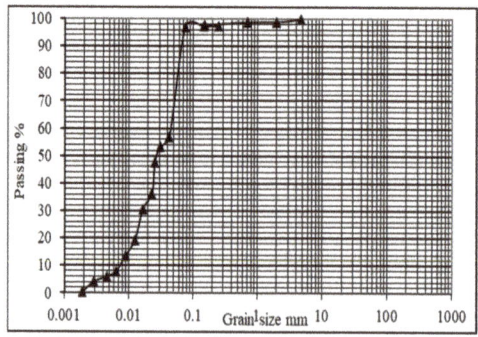

Figure 2. Gradation curve of back fill material.

Table 1. Properties of raw soil.

Test Description	Result	Test Description	Result
Specific gravity	2.76	Liquid limit	24.08%
OMC	14.50%	In situ Density	78.03 lb/ft^3
Natural moisture content	15–21%	MDD	110.48 lb/ft^2
Plastic limit	19.08%	UCS	983.1 lb/ft^2

4.2. Standard Proctor and Plate Load Test

The maximum dry density of the raw soil sample was measured as 110.48 lb/ft^3, while the optimum moisture content was found to be 15%. These values are depicted in Figure 3. The plate load test was conducted to determine the bearing capacity of the soil in different conditions: loose soil, partially compacted soil, and well-compacted soil after adding PET bottle strips at compositions of 0.4%, 0.7%, and 1.0% by weight. The stress settlement curve of loose soil is shown in Figure 4. The results showed that the bearing capacity increased with the addition of plastic strips up to a composition of 0.7%, as shown in Figure 5.

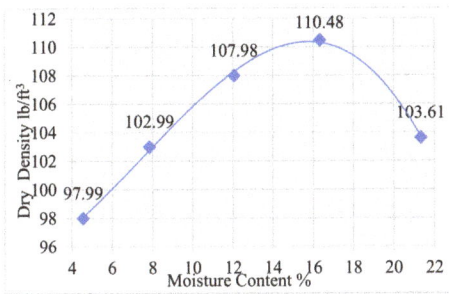

Figure 3. Moisture density curve.

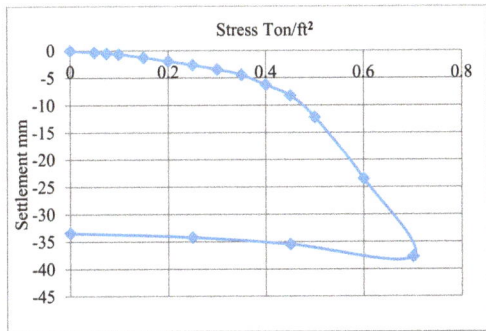

Figure 4. Stress settlement loose soil.

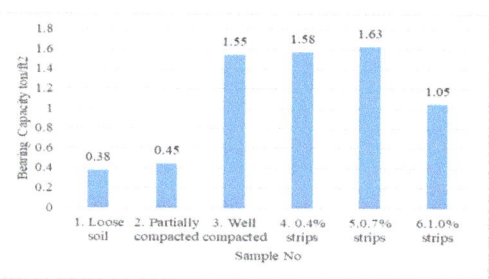

Figure 5. Effect of percentage PET bottle strips on bearing capacity.

5. Conclusions

The current study aimed to improve the bearing capacity of loosely filled fine-grained soils, hence reducing post-construction settlement. The uncontrolled filling comprised the silty clay group (CL-ML), as classified according to the Unified Soil Classification System (USCS). The following conclusions have been drawn following the laboratory and field investigation.

- The in situ density of the soil was 78.03 lb/ft^3, and the bearing capacity of the loose soil was 0.38 ton/ft^2.
- The dry density of well-compacted soil without any strips was found to be 94.88 lb/ft^3, and the corresponding bearing capacity was determined to be 1.55 ton/ft^2.
- The dry density of well-compacted soil with a 0.7% strip content was found to be 106.11 lb/ft^3, while the corresponding bearing capacity was determined to be 1.63 ton/ft^2.
- The maximum improvement in the bearing capacity of soil was observed as 328% compared to loosely filled soil in the field.

PET bottle strips are effective for soil stabilization and improving the engineering properties of soil. Utilizing PET bottle waste materials for soil stabilization is not only environmentally friendly but also economically viable.

Author Contributions: M.U.A.: Conceptualization, methodology, Supervision, writing—review and editing, administration; T.S.: Experimentation, formal analysis, investigation, writing—original draft preparation. All authors have read and agreed to the published version of the manuscript.

Funding: This research received no external funding.

Institutional Review Board Statement: Not applicable.

Informed Consent Statement: Not applicable.

Data Availability Statement: All the data has been incorporated in the manuscript.

Conflicts of Interest: The authors declare no conflict of interest.

References

1. Hassan, S.; Sharma, N. Strength improvement of clayey soil with waste plastic strips and cement kiln dust. *Int. J. Eng. Appl. Sci. Technol.* **2019**, *4*, 123–127. [CrossRef]
2. Arshid, M.U.; Sadia, Z.; Abdul, W. Evaluation of Cement Kiln Dust (Ckd) for the Improvement of the Thal Desert Dune Sand of Pakistan. In Proceedings of the 16th International Conference on Geotechnical Engineers, Lahore, Pakistan, 7–8 December 2022; Volume 16, pp. 131–135.
3. Waheed, A.; Arshid, M.; Khalid, R.A.; Gardezi, D.S.S.S. Soil Improvement Using Waste Marble Dust for Sustainable development. *Civ. Eng. J.* **2021**, *7*, 1594–1607. [CrossRef]
4. Kumar, T.; Panda, S.; Hameed, S.; Maity, J. Behaviour of soil by mixing of plastic strips. *Int. Res. J. Eng. Technol.* **2018**, *5*, 2578–2581.
5. Iravanian, A.; Haider, A.B. Soil Stabilization Using Waste Plastic Bottles Fibers: A Review Paper. *IOP Conf. Series Earth Environ. Sci.* **2020**, *614*. [CrossRef]

6. Roustaei, M.; Tavana, J.; Bayat, M. Influence of adding waste polyethylene terephthalate plastic strips on uniaxial compressive and tensile strength of cohesive soil. *Geopersia* **2022**, *12*, 39–51.
7. Jain, A.; Mittal, S.; Shukla, S.K. Use of polyethylene terephthalate fibres for mitigating the liquefaction-induced failures. *Geotext. Geomembranes* **2023**, *51*, 245–258. [CrossRef]

Disclaimer/Publisher's Note: The statements, opinions and data contained in all publications are solely those of the individual author(s) and contributor(s) and not of MDPI and/or the editor(s). MDPI and/or the editor(s) disclaim responsibility for any injury to people or property resulting from any ideas, methods, instructions or products referred to in the content.

Proceeding Paper

Framework for Energy Performance Measurement of Residential Buildings Considering Occupants' Energy Use Behavior [†]

Nida Azhar [1], Farrukh Arif [2,3,*] and Abdul Basit Khan [3]

1. Department of Urban and Infrastructure Engineering, NED University of Engineering and Technology, Karachi 75270, Pakistan; nazhar@cloud.neduet.edu.pk
2. Department of Civil Engineering, NED University of Engineering and Technology, Karachi 75270, Pakistan
3. NED University VR Center, NED University of Engineering and Technology, Karachi 75270, Pakistan; khan.pg3700916@cloud.neduet.edu.pk
* Correspondence: farrukh@cloud.neduet.edu.pk
† Presented at the 5th Conference on Sustainability in Civil Engineering (CSCE), Online, 3 August 2023.

Abstract: Buildings' contribution to global final energy use is about 30%, which makes them a primary focus for implementing energy-efficient measures. Building energy efficiency is an important consideration for residential buildings due to the significant environmental impact of energy consumption and the rising cost of energy. Estimating and optimizing a building's energy performance is an efficient method to reduce its environmental impact and cost. There exists a lack of accuracy in estimating the energy performance of a building due to approximations in the monitored data as well as a lack of consideration for occupants' energy use behavior. This study aimed to develop a comprehensive framework that assists in accurately estimating building energy performance considering occupants' energy use behavior. The framework proposed a scheme to collect occupant behavior data, such as occupancy patterns, appliance usage, and lighting conditions, through a living-lab setup and developing an occupants' behavior model that was utilized for more accurate building energy modeling and performance analysis.

Keywords: building energy performance; occupants behavior modeling; living-lab concept

1. Introduction

Buildings' contribution to global final energy use is about 30%, which makes them a primary focus for implementing energy-efficient measures [1]. A significant portion of this energy is wasted due to inappropriate building envelope design and construction. Building energy performance measurements can serve as a basis for building owners to make informed decisions for enhancing building energy efficiency. There is a growing concern in the building industry about the gap between the projected energy performance and the actual energy performance of buildings [2]. Bridging this performance gap is crucial in achieving the goal of reducing energy demand and enhancing building energy efficiency. The difference between the predicted and actual energy performance is due to approximations in the data as well as a lack of consideration for occupants' energy use behavior [3]. Therefore, a comprehensive energy performance measurement framework can help to effectively assess and quantify the building's energy efficiency.

The term occupant's behavior refers to the actions and responses exhibited by individuals within the building related to energy use and comfort, which are influenced by factors that include climate, building envelope, building energy and services systems, indoor and outdoor environments, time of the day, occupants' age and gender, and physiological, psychological, social, and economic factors [4].

2. Literature Review

To address the complex nature of occupant behavior and its impact on building energy consumption, Sun and Hong [5] proposed a framework for quantifying the influence of occupants' behavior on the energy savings achieved through energy conservation measures. Meanwhile, Wang et al. presented quantitative energy performance assessment methods specifically tailored for existing buildings, considering occupants' behavior and all other relevant factors to evaluate the buildings' energy efficiency [6]. Furthermore, Balvedi et al. [7] conducted a comprehensive review of various approaches and strategies available for gathering data on occupant behavior. They explored how these methods can be integrated into building energy simulation tools by incorporating occupant behavior models. It is widely observed that incorporating actual data for occupant behavior in energy analysis yields more accurate results as compared to energy simulations run without considering it [8].

The International Energy Agency identified six parameters that affect energy use in buildings. These parameters include climate, building envelope, building energy and services systems, indoor design criteria, building operation and maintenance, and occupant behavior. Each of these parameters plays a critical role in determining the energy efficiency of residential buildings and strategies for improving energy efficiency must consider each of these factors [9]. In addition, Chen et al. [10] in their study emphasized the need for a holistic approach to measuring the energy performance of residential buildings that considers all relevant factors and their potential impacts on energy consumption.

Laaroussi et al. [11] in their study identified the major issues and key drivers affecting occupants' behavior through an evaluation of existing approaches and methods for occupant behavior analysis. Furthermore, this study proposed and developed different methods to assess and predict the energy use behavior of occupants with better accuracy where conventional techniques such as structured and unstructured interviews, questionnaires, etc., prove to be inadequate. The study also emphasized integrating energy feedback programs into the building energy performance processes.

Incorporating an occupants' behavior model in the framework provides a more realistic evaluation of energy consumption patterns and assists in providing valuable insights into the factors affecting the energy performance of residential houses. Chen et al. [12] reviewed the impacts of occupant behavior on building energy consumption and established that the actual occupancy and the interactions with buildings are the key influencing factors determining the building energy consumption.

3. Methodology

This research study proposed a framework to measure the energy performance of residential buildings that incorporates occupants' energy use behavior with the purpose of accurately quantifying the impact of occupants' behavior on energy consumption in residential houses. The framework consists of three steps. (1) energy audit and data collection; (2) occupant behavior modeling; and (3) building energy modeling and performance analysis.

3.1. Energy Audit and Data Collection

The energy audit and data collection step begins with an assessment of the residential buildings, identifying the key influencing factors that affect energy consumption in residential buildings. For the energy audit and data collection, a hybrid method can be adopted that involves surveys, interviews, and on-site measurements using instruments to identify energy consumption and building envelope parameters in residential buildings, and installation of sensors for real-time monitoring of thermal properties of a building, indoor and outdoor environmental parameters, and occupants' energy use behavior through the living-lab concept [13].

The living-lab concept is a research methodology that focuses on the needs of end-users and stakeholders in the development of complex solutions. It involves creating a real-life test environment to sense, prototype, validate, and refine innovative solutions that address specific challenges [14].

3.2. Occupant Behavior Modeling

Occupant behavior modeling plays an essential role in understanding and predicting the energy consumption patterns of the occupants in a residential building. In a post-occupancy evaluation analysis [15], it was observed that occupant behaviors, including dissimilar presence at home, diverse occupancy levels, and differences in the occupants' thermal preferences play key roles in actual energy consumptions. Occupant behavior modeling involves developing mathematical models that incorporate various factors such as occupancy, interactions with the building systems, and occupants' preferences. The data collected in the energy audit and data collection step can be used to develop the mathematical model. Occupants' behavior is complex and diversified and has a stochastic nature rather than a deterministic one [16]. Therefore, stochastic occupant behavior model can be developed to capture the complex and diversified energy use behavior of occupants and generate synthetic occupancy schedules and occupants' energy use patterns with more precision over time. Such a model can be used to generate occupancy schedules and occupants' energy use patterns by simulating and predicting future states based on the current state and transition probabilities which can then be incorporated into the building energy simulation tools.

3.3. Building Energy Modeling and Performance Analysis

Building energy modeling and performance analysis begins with the development of a detailed energy model of the residential building, considering its physical characteristics, such as building geometry and orientation, building materials, insulation, HVAC systems, lighting, and appliances, using the data obtained from the energy audit and data collection. The real energy consumption data and the occupants' energy use behavior data such as occupancy schedules and occupants' energy use patterns can be incorporated into the energy model of the residential building. An energy simulation tool can then be used to simulate the energy performance of a residential building, with energy use intensity (EUI) serving as a metric to measure its energy consumption.

4. Building Energy Performance Measurement Framework

The framework depicted in Figure 1 comprises three main steps. The first step, energy audit and data collection, shall be carried out by collecting data related to building energy consumption and energy use; this includes; the data related to building envelope components and parameters such as building orientation, walls, roofing system, windows and glazing, doors, foundation and basement, exterior cladding, roof and window overhangs, solar heat gain coefficient (SHGC), insulation, R-Value and U-value, visual transmittance, etc., and monitoring occupants' energy use behavior through questionnaires, survey, interviews, and real-time monitoring through IOT sensors and data-logging sensors. The second step is to develop an occupant behavior model to generate synthetic occupancy schedules and energy use patterns using the data collected through occupant behavior monitoring. The data collected in the first and second step is then analyzed to perform building energy modeling and performance analysis.

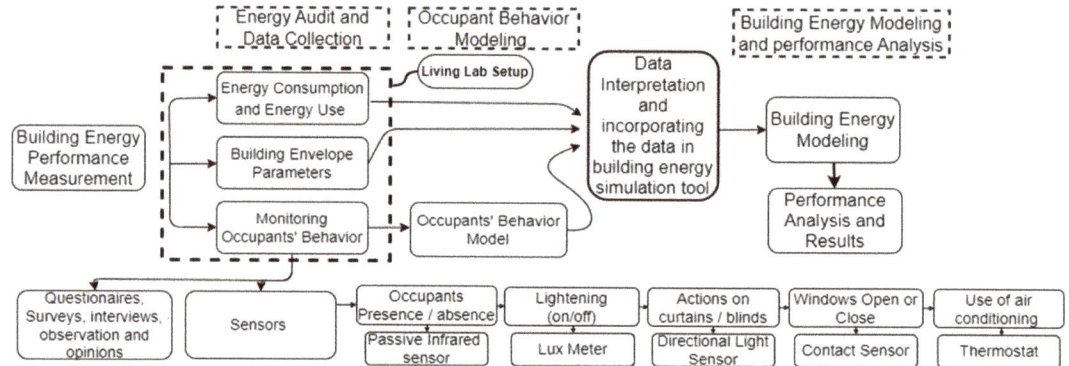

Figure 1. Building energy performance measurement framework.

5. Conclusions

This framework integrates an occupant behavior model, which can capture the complex and diversified energy use behavior of occupants. Therefore, this framework facilitates more accurate measurements of energy performance based on real-time data of the occupants' behavior and can enable the evaluation of different energy-saving strategies and the development of more efficient building designs customized to the occupant's behavior.

Author Contributions: Conceptualization, N.A. and F.A.; methodology, N.A.; writing—original draft preparation, A.B.K.; writing—review and editing, F.A.; visualization, N.A. and A.B.K.; supervision, F.A.; project administration, F.A.; funding acquisition, F.A. and N.A. All authors have read and agreed to the published version of the manuscript.

Funding: This research is a part of Sindh Higher Education Commission project-168 "Development of Energy Efficient Housing Design for Karachi using Living Lab-Virtual Reality Integration." The APC was funded through same project as well.

Institutional Review Board Statement: Not applicable.

Informed Consent Statement: Not applicable.

Data Availability Statement: All the relevant data has been included in the paper.

Conflicts of Interest: The authors declare no conflict of interest.

References

1. International Energy Agency. Buildings—Analysis—IEA. 2022. Available online: https://www.iea.org/reports/buildings (accessed on 15 March 2023).
2. De Wilde, P. The gap between predicted and measured energy performance of buildings: A framework for investigation. *Autom. Constr.* **2014**, *41*, 40–49. [CrossRef]
3. Paone, A.; Bacher, J.P. The impact of building occupant behavior on energy efficiency and methods to influence it: A review of the state of the art. *Energies* **2018**, *11*, 953. [CrossRef]
4. Casini, M. *Construction 4.0: Advanced Technology, Tools and Materials for the Digital Transformation of the Construction Industry*; Woodhead Publishing: Cambridge, UK, 2021.
5. Sun, K.; Hong, T.; Kim, J.H. A simulation framework for quantifying the influence of occupant behavior on savings of energy efficiency measures. *Build. Simul. Conf. Proc.* **2017**, *3*, 1646–1654. [CrossRef]
6. Wang, S.; Yan, C.; Xiao, F. Quantitative energy performance assessment methods for existing buildings. *Energy Build.* **2012**, *55*, 873–888. [CrossRef]
7. Balvedi, B.F.; Ghisi, E.; Lamberts, R. A review of occupant behaviour in residential buildings. *Energy Build.* **2018**, *174*, 495–505. [CrossRef]
8. Eguaras-Martínez, M.; Vidaurre-Arbizu, M.; Martín-Gómez, C. Simulation and evaluation of building information modeling in a real pilot site. *Appl. Energy* **2014**, *114*, 475–484. [CrossRef]
9. Yoshino, H.; Hong, T.; Nord, N. IEA EBC annex 53: Total energy use in buildings—Analysis and evaluation methods. *Energy Build.* **2017**, *152*, 124–136. [CrossRef]

10. Chen, S.; Zhang, G.; Xia, X.; Setunge, S.; Shi, L. A review of internal and external influencing factors on energy efficiency design of buildings. *Energy Build.* **2020**, *216*, 109944. [CrossRef]
11. Laaroussi, Y.; Bahrar, M.; El Mankibi, M.; Draoui, A.; Si-Larbi, A. Occupant presence and behavior: A major issue for building energy performance simulation and assessment. *Sustain. Cities Soc.* **2020**, *63*, 102420. [CrossRef]
12. Chen, S.; Zhang, G.; Xia, X.; Chen, Y.; Setunge, S.; Shi, L. The impacts of occupant behavior on building energy consumption: A review. *Sustain. Energy Technol. Assess.* **2021**, *45*, 101212. [CrossRef]
13. Giannouli, I.; Tourkolias, C.; Zuidema, C.; Tasopoulou, A.; Blathra, S.; Salemink, K.; Gugerell, K.; Georgiou, P.; Chalatsis, T.; Christidou, C.; et al. A methodological approach for holistic energy planning using the living lab concept: The case of the prefecture of karditsa. *Eur. J. Environ. Sci.* **2018**, *8*, 14–22. [CrossRef]
14. Arif, F.; Zaman, M.; Khalid, R. Living Lab Concept for Sensor Based Energy Performance Assessment of Houses. *Neutron* **2022**, *21*, 105–111. [CrossRef]
15. Zero Carbon HUB. *Post-Occupancy Evaluation, Rowner Research Project Phase Two*; Zero Carbon HUB: Milton Keynes, UK, 2015.
16. Wang, C.; Yan, D.; Jiang, Y. A novel approach for building occupancy simulation. *Build. Simul.* **2011**, *4*, 149–167. [CrossRef]

Disclaimer/Publisher's Note: The statements, opinions and data contained in all publications are solely those of the individual author(s) and contributor(s) and not of MDPI and/or the editor(s). MDPI and/or the editor(s) disclaim responsibility for any injury to people or property resulting from any ideas, methods, instructions or products referred to in the content.

Proceeding Paper

The Behavior of Retrofitted GPC Columns under Eccentric Loading [†]

Shahzaib Farooq *, Faheem Butt and Rana Muhammad Waqas

Department of Civil Engineering, University of Engineering and Technology, Taxila 47080, Pakistan; faheem.butt@uettaxila.edu.pk (F.B.); rana.waqas@uettaxila.edu.pk (R.M.W.)
* Correspondence: shahzaibfarooq010@gmail.com
† Presented at the 5th Conference on Sustainability in Civil Engineering (CSCE), Online, 3 August 2023.

Abstract: Geopolymer concrete (GPC) has been the subject of ongoing research as a suitable substitute for conventional concrete production because of its benefits for the environment. However, there is little research regarding retrofitting the structural part if a GPC member fails. The current study thus concentrates on the damaged GPC structural members/columns. For this purpose, twelve columns which include four CC columns, four GPC Columns, and four FRGPC columns, were retrofitted with CFRP sheets and tested in the electrohydraulic testing apparatus (5000 kN). The results showed significant improvement in the ultimate load value of all 12 columns. Axial strain in all 12 columns also increased significantly. The ductility index of the columns was also calculated using axial strain values. The axial load–displacement behavior, ductility, and loading capacity of the evaluated columns are all significantly improved by the addition of steel fibers.

Keywords: carbon fiber reinforced polymer (CFRP); eccentricity; fiber reinforced geopolymer concrete (FRGPC); geopolymer concrete (GPC)

1. Introduction

Geopolymer concrete (GPC) has evolved as a new option that may completely eliminate the need for cement while promoting the efficient use of waste materials. However, if a GPC structural member fails, there is little study about the retrofitting of that member. Therefore, the present study focuses on the damaged GPC structural columns. Fiber-reinforced polymer (FRP) composites, the newest modern composite materials, have recently surpassed conventional retrofitting techniques in demand. FRP jackets are the ideal material because of their high rigidity and high strength-to-weight ratio. As a result, FRP has seen significant application in retrofitting.

Numerous tests and theoretical analyses have clearly proved that wrapping FRP composites around columns is a very successful approach. Yang et al. investigated the eccentric compression loading of rectangular high-strength concrete columns restricted with carbon fiber-reinforced polymer (CFRP) [1]. Zeng et al. investigated the cyclic axial compression behavior of FRP spiral strip-confined concrete [2]. Askandar et al. examined the behavior of RC beams reinforced with FRP strips under the combined action of torsion and bending [3]. For FRP spiral strip-confined concrete, Liao et al. researched the stress–strain behavior and design-oriented model [4]. The partially FRP strengthening approach is a viable option, particularly for columns that require moderate increases in strength and deformation capacity [5].

In light of the previously mentioned, the purpose of this research is to explore the axial compressive behavior of partially FRP confined Fiber Reinforced Geopolymer Concrete (FRGC). The current investigation involved the retrofitting and testing of 12 columns using CFRP.

2. Experimental Procedure

A total of 12 columns with cross-sections of 200 mm square and heights of 1000 mm were examined in the present study. Six deformed 12 mm diameter bars were used to brace the columns longitudinally. In all instances, transverse reinforcement was supplied as closed ties of diameter 6 mm bars spaced 100 mm from centers. Deformed steel with a yield strength of 300 MPa for D6 (6 mm) bars and 450 MPa for D12 (12 mm) bars was employed. CFRP wrapping of the columns was carried out using a unique pattern, as shown in Figure 1a. The results are compared between CC columns and GPC columns. In order to evaluate the columns, two loading scenarios were used, concentric loading and eccentric loading with varying eccentricities (eccentricity e = 15, 35, and 50 mm). Sample details are given in Table 1.

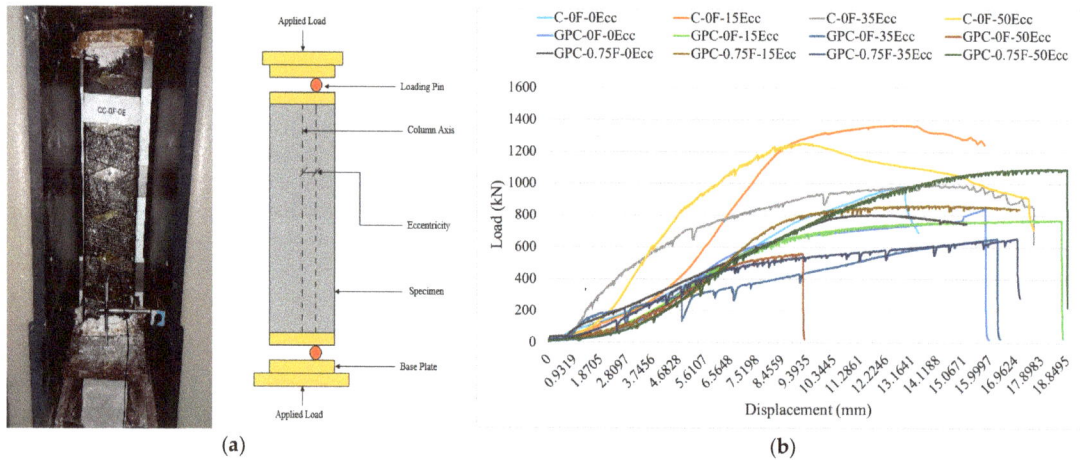

Figure 1. (a) Load mechanism for CFRP columns; (b) ultimate load of CC and GPC columns.

Table 1. Mix proportion and material quantities of mixes [6].

Sr. No.	Group ID	Specimen ID	Mix Proportion					Mix Quantities (kg/m³)						
			OPC	FA	SG	Sand	CA	OPC	FA	SG	NaOH	Na₂SiO₃	SP	Water
1	CC	C-0F-0Ecc	100%	-	-	640	1201	370	-	-	53	107	4	170
		C-0F-15Ecc	100%	-	-	643	1206	370	-	-	53	107	4	170
		C-0F-35Ecc	100%	-	-	640	1201	370	-	-	53	107	4	170
		C-0F-50Ecc	100%	-	-	643	1206	370	-	-	53	107	4	170
2	GPC	GPC-0F-0Ecc	-	50%	50%	643	1206	-	200	200	53	107	8	-
		GPC-0F-15Ecc	-	50%	50%	643	1206	-	200	200	53	107	8	-
		GPC-0F-35Ecc	-	50%	50%	646	1212	-	200	200	53	107	8	-
		GPC-0F-50Ecc	-	50%	50%	643	1206	-	200	200	53	107	8	-
		GPC-0.75F-0Ecc	-	50%	50%	643	1206	-	200	200	53	107	12	-
		GPC-0.75F-15Ecc	-	50%	50%	644	1208	-	200	200	53	107	12	-
		GPC-0.75F-35Ecc	-	50%	50%	643	1206	-	200	200	53	107	12	-
		GPC-0.75F-50Ecc	-	50%	50%	647	1214	-	200	200	53	107	12	-

2.1. Preparation of Specimen

Firstly, the repairing of the specimens is carried out using geopolymer mortar having 50% fly ash and 50% slag as a binder. The specimens were then dried in atmospheric conditions for 28 days. Secondly, retrofitting of the specimens was carried out using CFRP sheets 3 mm thick having a width of 82 mm. A total of four CFRP sheets were used for wrapping each specimen, two clockwise and two anti-clockwise, at an angle of 20°. The CFRP should be placed firmly against the GPC and CC surface in order to make good contact and remove any air pockets between it and the concrete surface.

2.2. Testing

At 28 days, unidirectional axial loading was given to each specimen. A 5000 kN capacity electrohydraulic testing equipment was used to apply the loading. Under displacement control conditions, the columns were tested to the point of failure. The load was applied at regular 1 mm/s intervals. For the concentrically loaded columns, a similar system was employed, but there was no loading pin. In order to prevent columns from failing prematurely due to overstressing, steel collars of thickness 3.2 mm having a width of 76 mm were attached at both ends of each column prior to testing. On the top and bottom sides of the columns, a thin coating of Plaster of Paris was also used to provide a level surface for the test's uniform weight distribution. A magnetic Linear Variable Differential Transformer (LVDT) of 20 mm capacity and 0.001 mm accuracy was vertically aligned with the base plate of the machine to measure the axial deformation in the specimen.

3. Research Methodology

First, the surface of the specimen was prepared, and any loose or broken material was removed. Second, a geopolymer mortar with a binder made of 50% fly ash and 50% slag is used to repair the specimens. Wrapping of CFRP strips was carried out around the column after applying the bonding agent to the finished surface of the column, make sure the CFRP strips were at a 20-degree angle with the column's horizontal axis. Use a 20 mm capacity magnetic LVDT and electrohydraulic testing apparatus (5000 kN), which can show the structural performance and behavior of the columns. The LVDT is accurately positioned and calibrated in order to precisely measure the vertical deflection of the column during the test. As we gradually added force to the column until it reached its full capacity, we set the deflection rate to 1 mm per minute. We then took the deflection data that the LVDT computer supplied to chart the behavior of the column as the load increased. We repeated the test until the column failed or started to distort visibly.

4. Results

The ultimate load of the specimens obtained from the experimental results is shown in Table 2. We can clearly see that the specimens from the GPC group showed lower ultimate load values than those from the CC group. The ultimate load of CC, GPC, and FRGPC columns was improved from previous results. The ultimate load values after retrofitting shows significant improvement. The fact that the ultimate load of the FRGPC column on concentric loading is lower than the previous value is due to the rusting of steel fibers present in the specimen. Figure 1b demonstrates how changing the value of eccentricity in the tested specimens affects load levels. In other words, as the eccentricity of the axial load increases, the ability of the column to carry loads decreases, which is linked to its eccentricity. The ductility index for all specimens was also calculated.

Table 2. Columns axial strength and ductility index.

Sr. No.	Group ID	Specimen ID	Quantity of Fiber (%)	Eccentricity	Before Retrofitting P_{max} (kN)	After Retrofitting P'_{max} (kN)	Axial Deformation at P'_{max} (mm)	Ductility Index
1	CC	C-0F-0Ecc	0	0	945	980.498	12.741	1.01
		C-0F-15Ecc	0	15	712	1366.153	13.043	1.21
		C-0F-35Ecc	0	35	430	994.696	13.485	1.35
		C-0F-50Ecc	0	50	330	1253.35	9.197	1.89
2	GPC	GPC-0F-0Ecc	0	0	860	844.977	14.959	1.00
		GPC-0F-15Ecc	0	15	570	771.793	17.505	1.07
		GPC-0F-35Ecc	0	35	335	654.541	16.241	1.01
		GPC-0F-50Ecc	0	50	249	559.063	9.133	1.01
		GPC-0.75F-0Ecc	75%	0	1000	801.639	11.621	1.30
		GPC-0.75F-15Ecc	75%	15	720	861.009	13.741	1.24
		GPC-0.75F-35Ecc	75%	35	460	657.994	17.052	1.01
		GPC-0.75F-50Ecc	75%	50	340	1095.335	18.170	1.04

5. Conclusions

The main focus of this research is to provide the proper solution for retrofitting GPC columns. From all the above research, we can now say that CFRP wrapping of GPC columns is a violable solution. This paper presents the results of twelve columns, including four reference columns, four GPC columns, and four FRGC columns. All 12 columns were wrapped with CFRP sheets in a particular manner. The ultimate strength of the CC columns, GPC columns, and FRGPC columns was compared to a previous study due to retrofitting. Considering the experimental and theoretical findings in this research, the ultimate load values of CC columns, GPC columns, and FRGPC columns increased from the values that were obtained from previous studies. The axial displacement of all the columns also significantly improved. The ductility index of all 12 columns also increased.

Author Contributions: Conceptualization, S.F. and F.B.; methodology, R.M.W.; software, S.F.; validation, S.F.; formal analysis, S.F.; investigation, F.B.; resources, S.F.; data curation, R.M.W.; writing—original draft preparation, S.F.; writing—review and editing, S.F.; visualization, F.B.; supervision, F.B.; project administration, F.B.; funding acquisition, S.F. All authors have read and agreed to the published version of the manuscript.

Funding: This research received no external funding.

Institutional Review Board Statement: Not applicable.

Informed Consent Statement: Not applicable.

Data Availability Statement: The data presented in this study are available on request from the corresponding author. The data are not publicly available due to confidentiality purposes.

Conflicts of Interest: The authors declare no conflict of interest.

References

1. Yang, J.; Wang, J.; Wang, Z. Rectangular high-strength concrete columns confined with carbon fiber-reinforced polymer (CFRP) under eccentric compression loading. *Constr. Build. Mater.* **2018**, *193*, 604–622. [CrossRef]
2. Zeng, J.J.; Liao, J.; Ye, Y.Y.; Guo, Y.C.; Zheng, Y.; Tan, L.H. Behavior of FRP spiral strip-confined concrete under cyclic axial compression. *Constr. Build. Mater.* **2021**, *295*, 123544. [CrossRef]
3. Askandar, N.H.; Mahmood, A.D.; Kurda, R. Behaviour of RC beams strengthened with FRP strips under combined action of torsion and bending. *Eur. J. Environ. Civ. Eng.* **2022**, *26*, 4263–4279. [CrossRef]
4. Liao, J.; Zeng, J.J.; Jiang, C.; Li, J.X.; Yuan, J.S. Stress-strain behavior and design-oriented model for FRP spiral strip-confined concrete. *Compos. Struct.* **2022**, *293*, 115747. [CrossRef]
5. Zeng, J.J.; Guo, Y.C.; Gao, W.Y.; Li, J.Z.; Xie, J.H. Behavior of partially and fully FRP-confined circularized square columns under axial compression. *Constr. Build. Mater.* **2017**, *152*, 319–332. [CrossRef]
6. Waqas, R.M.; Butt, F. Behavior of Quarry Rock Dust, Fly Ash and Slag Based Geopolymer Concrete Columns Reinforced with Steel Fibers under Eccentric Loading. *Appl. Sci.* **2021**, *11*, 6740. [CrossRef]

Disclaimer/Publisher's Note: The statements, opinions and data contained in all publications are solely those of the individual author(s) and contributor(s) and not of MDPI and/or the editor(s). MDPI and/or the editor(s) disclaim responsibility for any injury to people or property resulting from any ideas, methods, instructions or products referred to in the content.

Proceeding Paper

Composite Fibers in Concrete: Properties, Challenges, and Future Directions [†]

Samiullah Khan [1], Safeer Khattak [1,*] and Hamza Khan [2]

[1] Department of Civil Engineering, Comsats University, Campus Abbottabad, Islamabad 22001, Pakistan; 2017civ149@student.uet.edu.pk

[2] School of Civil and Environmental Engineering, National University of Science and Technology, H/12 Campus, Islamabad 44000, Pakistan; hkhan.ms22nice@student.nust.edu.pk

* Correspondence: civ405416@gmail.com; Tel.: +92-3094718924

[†] Presented at the 5th Conference on Sustainability in Civil Engineering (CSCE), Online, 3 August 2023.

Abstract: Composite fibers are an essential component of modern concrete structures, providing enhanced mechanical properties and durability. There are different types of fibers commonly used in concrete such as steel, glass, carbon, polypropylene fibers, etc. Composite fibers include a combination of two or more of these fibers. Composite fibers are discussed in detail, along with their properties and benefits. This paper also highlights the effects of composite fibers on the various properties of concrete, such as compressive strength, tensile strength, toughness, and durability. Furthermore, this paper discusses some of the challenges and limitations associated with the use of composite fibers in concrete, including issues related to fiber dispersion, fiber–matrix interactions, and cost-effectiveness. Finally, this paper concludes with a discussion of the future directions of research on composite fibers in concrete, focusing on potential advancements in fiber technology, improved manufacturing techniques, and the development of new fiber–matrix systems. Overall, this paper provides a comprehensive overview of the current state of the art in composite fibers in concrete and serves as a valuable resource for researchers and practitioners in the field.

Keywords: fiber composites; concrete matrix; tensile strength; tensile stress; durability; mechanical properties

1. Introduction

Due to its adaptability, toughness, and strength, concrete is the most widely used building material worldwide. However, it is prone to cracking and failure under tensile stress, which can result in structural damage and a shorter service life despite its many benefits [1]. Different kinds of fibers are added into concrete to improve its mechanical properties, such as its strength, hardness, and durability, in order to address this problem and increase its tensile strength. Modern concrete constructions require the use of these composite fibers, which are created from materials like steel, glass, carbon, and polypropylene and enable the structures to resist significant stress and strain. The development of fibers dates back to the Neolithic period when people used wheat straw in mud to increase its tensile strength and employed it for various purposes. Over the past few decades, the usage of fibers in concrete has become more widespread. For instance, steel fibers have been utilized to reinforce concrete and lessen cracking since the 1960s [2,3]. Additionally, the use of glass and carbon fibers has become more popular in concrete construction, which offers good corrosion resistance and a high strength-to-weight ratio [4].

This paper provides a thorough analysis of the most recent, cutting-edge research on composite fibers in concrete. The paper begins by outlining the many kinds of fibers that are frequently used in concrete, along with their characteristics and advantages. The impacts of composite fibers on different concrete parameters, including compressive strength, tensile strength, toughness, and durability, are then examined [5]. This research also

outlines a number of difficulties and restrictions related to the usage of composite fibers in concrete, including issues related to fiber distribution, fiber–matrix interactions, and cost-effectiveness [6,7]. The use of composite fibers in concrete is then illustrated with examples from both structural and non-structural situations. Finally, the paper concludes with a discussion of future directions in concrete composite fiber research, with a focus on potential advances in fiber technology, improvements in manufacturing techniques, and the development of new fiber matrix systems. Overall, the purpose of this study is to introduce the readers to the field of fibers and their applications while also presenting a thorough grasp of composite fibers in concrete.

2. Types of Composite Fibers in Concrete

Composite fibers are typically made by combining two or more types of fibers to create a material that exhibits the desired properties. The utilization of fibers to enhance the mechanical properties of concrete is quite common today [8]. By using different types of fibers in concrete, its properties can be altered to the desired strength and durability [9]. Table 1 presents a tabular comparison between different types of composite fibers along with their properties and composition.

Table 1. Different types of composite fibers, their properties, and their composition.

Composite Fiber Type	Composition	Properties
Carbon Fiber Composite	Carbon fibers + polymer matrix	High strength-to-weight ratio, excellent stiffness, fatigue resistance
Glass Fiber Composites	Glass fibers + polymer matrix	Good strength + stiffness, low cost
Basalt Fiber Composites	Basalt fibers + polymer matrix	Good tensile strength, excellent resistance to heat
Ceramic Fiber Composites	Ceramic fibers + polymer or metal matrix	Excellent thermal insulation, high-temperature resistance

3. Mechanical Properties of Concrete

Concrete's mechanical properties, including compressive strength, tensile strength, toughness, and durability, can be greatly improved by the addition of composite fibers. We present a general overview of how composite fibers affect certain characteristics. The inclusion of fibers in the concrete matrix affects the mechanical characteristics of fiber-reinforced concrete (FRC) [10]. Figure 1a,b presents values of compressive strength and tensile strength for different percentages of steel polymer matrix (0–1.25%). The first batch contains no fiber reinforcement and subsequent batches contain fibers in % increments of 0.25%.

3.1. Compressive Strength

By improving its ability to withstand external loads, the addition of composite fibers can increase the compressive strength of concrete. For instance, it was discovered that steel fibers can boost concrete's compressive strength by up to 15%, depending on the fiber volume fraction and aspect ratio, as illustrated in Figure 1a. The bridging effect of fibers, which slows the spread of cracks and improves material ductility, is thought to be responsible for the increase in compressive strength [7].

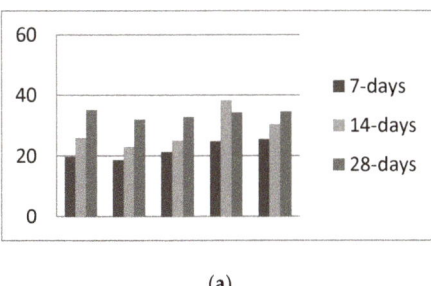 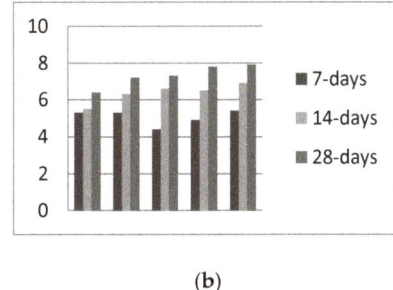

(a) (b)

Figure 1. Mechanical properties of fiber-reinforced concrete. (**a**) Compressive strength. (**b**) Flexural strength of composite fiber concrete.

3.2. Tensile Strength

One of the issues with concrete is its low tensile strength, and to address this issue, fiber-reinforced concrete, a new type of concrete composite, was designed [8]. Concrete's tensile strength can be considerably increased by the inclusion of composite fibers, making it less likely to crack or break under tensile stress. As seen in Figure 1b, polymer steel composites, for instance, have been proven to increase the tensile strength of concrete by up to 50%.

3.3. Ductility

The splitting tensile strength of fiber-reinforced concrete (FRC) is a measure of its resistance to tensile forces applied perpendicular to the direction of the applied load. It is a crucial factor in determining how well FRC performs overall in terms of crack resistance across a range of applications.

3.4. Modulus of Elasticity

The modulus of elasticity, which gauges a material's stiffness or rigidity, can be significantly impacted by the addition of fibers to concrete. The mechanical properties of concrete change as a result of the fibers, which act as a reinforcement within the concrete matrix [3]. When fibers are added to concrete, the material becomes more composite by bridging fissures and has a higher load-bearing capability.

3.5. Toughness

The ability of a material to withstand energy absorption and prevent fracture or crack propagation is referred to as toughness. By bridging and stopping cracks that may occur while the concrete is under load, composite fibers help to increase the concrete's durability [8]. Composite fibers serve as reinforcement when cracks first appear in the concrete matrix, dispersing stress and stopping the cracks from spreading.

3.6. Durability

The ability of a substance to resist deterioration and degradation over time, particularly when subjected to extreme environmental conditions, is referred to as durability. Concrete's durability is increased by composite fibers in a number of ways.

4. Challenges and Limitations of Using FRC

Composite fibers offer many advantages when used as reinforcement in concrete, but there are also some challenges and limitations to consider. Cost is one of the main drawbacks of employing composite fibers in concrete [5]. The compatibility of composite fibers with the concrete matrix is another drawback. To ensure effective fiber adhesion to the matrix, the characteristics of the fibers and the concrete mix must be precisely matched. Composite reinforcements may be less effective and less durable due to incompatibility [5–7].

The use of composite fibers can also provide some difficulties. Composite fibers can present additional difficulties during the mixing procedure.

5. Future Directions in Research on Composite Fibers in Concrete

Recent years have seen a lot of interest in the use of composite fibers in concrete, and research is still being performed to examine potential new applications and difficulties. The use of short fibers or whiskers is both the simplest and most efficient method. It was discovered that this technique considerably increases concrete's tensile strength and decreases tensile cracking. To create new varieties of composite fibers with enhanced characteristics and performance compared to existing materials, research is currently being conducted. Advanced manufacturing techniques such as 3D printing and automated fiber placement enable composite fibers to be more accurately and efficiently incorporated into concrete [2–4]. These techniques can also be used to tailor the orientation and distribution of fibers within concrete, which can improve the overall performance of composite reinforcements.

6. Conclusions

Finally, this conference paper on composite fibers will offer insightful information about the wide variety of composite fibers and their uses. This article presents a comparative analysis of the various fiber types, including carbon fiber, glass fiber, steel fiber, natural fiber, basalt fiber, and ceramic fiber. A high strength-to-weight ratio, stiffness, fatigue resistance, impact resistance, heat resistance, affordability, a renewable nature, and biodegradability are just a few of the special advantages that each type of fiber provides. This paper's objective is to present a thorough overview of the various aspects of CFRPs, including their structural and durability properties, ongoing research into their use as reinforcing materials for concrete structures, and the application of FRP sheets in the rehabilitation of deteriorated concrete structures. Though they have not been thoroughly examined, topics pertaining to the production of FRP composites and design processes have been included for informational and comprehensive purposes.

Author Contributions: Conceptualization, S.K. (Samiullah Khan) and S.K. (Safeer Khattak); methodology, S.K. (Samiullah Khan); software, S.K. (Samiullah Khan); validation, S.K. (Samiullah Khan), H.K. and S.K. (Safeer Khattak); formal analysis, S.K. (Samiullah Khan); investigation, S.K. (Samiullah Khan); resources, S.K. (Safeer Khattak); data curation, H.K.; writing—original draft preparation, S.K. (Samiullah Khan). All authors have read and agreed to the published version of the manuscript.

Funding: This research received no external funding.

Institutional Review Board Statement: Not applicable.

Informed Consent Statement: Not applicable.

Data Availability Statement: All data is obtained from open source articles and journals.

Conflicts of Interest: The authors declare no conflict of interest.

References

1. Monfared, V.; Ramakrishna, S.; Alizadeh, A.; Hekmatifar, M. A systematic study on composite materials in civil engineering. *Ain Shams Eng. J.* **2023**, *6*, 102251. [CrossRef]
2. Yoo, D.-Y.; Banthia, N. Mechanical properties of ultra-high-performance fiber-reinforced concrete: A review. *Cem. Concr. Compos.* **2016**, *73*, 267–280. [CrossRef]
3. Fediuk, R.; Smoliakov, A.; Muraviov, A. Mechanical Properties of Fiber-Reinforced Concrete Using Composite Binders. *Adv. Mater. Sci. Eng.* **2017**, *2017*, 2316347. [CrossRef]
4. Mahmood, A.; Noman, M.T.; Pechočiaková, M.; Amor, N.; Petrů, M.; Abdelkader, M.; Militký, J.; Sozcu, S.; Hassan, S.Z.U. Geopolymers and Fiber-Reinforced Concrete Composites in Civil Engineering. *Polymers* **2021**, *13*, 2099. [CrossRef] [PubMed]
5. Mobini, M.H.; Khaloo, A.; Hosseini, P.; Esrafili, A. Mechanical properties of fiber-reinforced high-performance concrete incorporating pyrogenic nanosilica with different surface areas. *Constr. Build. Mater.* **2015**, *101*, 130–140. [CrossRef]
6. Bheel, N.; Awoyera, P.; Aluko, O.; Mahro, S.; Viloria, A.; Sierra, C.A.S. Sustainable composite development: Novel use of human hair as fiber in concrete. *Case Stud. Constr. Mater.* **2020**, *13*, e00412. [CrossRef]

7. Klyuev, S.V.; Khezhev, T.; Pukharenko, Y.; Klyuev, A. Fiber Concrete for Industrial and Civil Construction. *Mater. Sci. Forum* **2019**, *945*, 120–124. [CrossRef]
8. Lee, H.; Choi, M.K.; Kim, B.-J. Structural and functional properties of fiber reinforced concrete composites for construction applications. *J. Ind. Eng. Chem.* **2023**, *125*, 38–49. [CrossRef]
9. You, X.; Lin, L.; Fu, B.; Xiang, Y. Ultra-high performance concrete reinforced with macro fibres recycled from waste GFRP composites. *Case Stud. Constr. Mater.* **2023**, *18*, e02120. [CrossRef]
10. Marcalikova, Z.; Cajka, R.; Bilek, V.; Bujdos, D.; Sucharda, O. Determination of Mechanical Characteristics for Fiber-Reinforced Concrete with Straight and Hooked Fibers. *Crystals* **2020**, *10*, 545. [CrossRef]

Disclaimer/Publisher's Note: The statements, opinions and data contained in all publications are solely those of the individual author(s) and contributor(s) and not of MDPI and/or the editor(s). MDPI and/or the editor(s) disclaim responsibility for any injury to people or property resulting from any ideas, methods, instructions or products referred to in the content.

Proceeding Paper

Utilizing Corn Cob Ash and Bauxite as One-Part Geopolymer: A Sustainable Approach for Construction Materials [†]

Raheel Arif *, Ammar Iqtidar and Safeer Ullah Khattak

Department of Civil Engineering, COMSATS University Islamabad, Abbottabad Campus, Abbottabad 22020, Pakistan; ammariqtidar@gmail.com (A.I.); safeer@cuiatd.edu.pk (S.U.K.)
* Correspondence: raheelarif021@gmail.com
† Presented at the 5th Conference on Sustainability in Civil Engineering (CSCE), Online, 3 August 2023.

Abstract: This research focused on creating sustainable geopolymer mortar using waste materials such as corn cob ash (CCA) and bauxite. The CCA was obtained by burning corn cobs in an open environment and then further treated at 600 °C to remove carbon impurities. The resulting ash was ground to improve reactivity and used as a binder. Sodium silicate was used as an activator for geopolymerization. The geopolymer was prepared by combining the binder with fine aggregate, ground bauxite, and CCA in different proportions. The curing process involved heating the samples at 70 °C for 24 h followed by ambient temperature curing. Compression testing was conducted at 7, 14, and 28 days to assess the strength and durability of the geopolymer mortar. Testing was performed according to ASTM standards.

Keywords: geopolymer; corn cob ash; bauxite; construction; materials

1. Introduction

Concrete is the most used building material all around the world. Concrete is a composite material composed of fine and coarse aggregate bonded together with cement. Many researchers have tried to reduce the harmful environmental effects of cement by replacing it with supplementary cementations materials such as rice rusk ash, fly ash, ground granulated blast furnace slag, and corn cob ash (CCA) [1]. Pakistan produced almost 7.9 million tons of corn from 2021–2022 and this production is increasing every year at a rate of 4.97% [2]. One possible way of utilizing CCA is its use as secondary cementitious material (SCM). Studies have shown that CCA possesses all the properties which would make it suitable to be used as an SCM [3]. Bauxite is a naturally occurring rock which is rich in alumina. It is found in abundance in the Kotli AJK region. The purpose of using bauxite with CCA in this research is that it will not cause any sudden depletion of natural resources as no other industry utilizes it and it is considered waste. It is easily mineable as it is soft rock and contains 45–50% Al_2O_3, not more than 20% Fe_2O_3, and 3–5% Silica [4].

This research will be limited to only the synthesis of one-part geopolymer mortar (GM). Different mass ratios of CCA and bauxite will be used to find the optimum combination of CCA and bauxite that can be adopted for preparation of a high-strength geopolymer. The research aims to contribute to sustainable construction practices and explore the potential of agricultural waste materials in geopolymer technology.

2. Research Methodology

The research was conducted as per the following procedures and details:

2.1. Materials Preparation

Corn cob was burnt in an open environment before burning in a controlled environment at 600 °C for 2 h. The obtained ash was crushed in a jar mill and then passed through sieve No. 200. Bauxite was jar milled and passed through sieve No. 200.

2.2. Mortar Preparation

Mix design for GM is shown in Table 1 below:

Table 1. Mix design for GM.

Material	Percentage by Weight (%)
CCA	10, 20, 30 percent weight of binder
Bauxite	90, 80, 70 percent weight of binder
Sand	50
Activator (Sodium Silicate)	10% the weight of binder
Water-to-Binder Ratio	0.3
Super Plasticizer	2% the weight of binder

Cubes were cast for different tests according to mix design.

2.3. Curing

Curing was conducted at the high temperature of 70 °C in an oven for 7, 14, and 28 days.

2.4. Tests

Sieve analysis was conducted on fine aggregates (ASTM C136-05) [5]. Temperatures of mortar mixes were checked (ASTM C1064) [6]. Initial setting time of GM was noted through Vicat needle apparatus in accordance with ASTM Standard C191 [7]. Slump test was conducted on all the cubes of GM (ASTM C143) [8]. Compressive strength test was performed to check the compressive strength of GM in a universal testing machine (UTM) (ASTM C109) [9]. Cubes were observed for appearance of cracks.

3. Results

3.1. Sieve Analysis of Fine Aggregates

The sand was obtained from Lawrencepur quarry. Sieve analysis was conducted to determine the fineness modulus of the sand. Equation (1) was used to find the fineness modulus. Fineness modulus helps in determining the particle size of sand. It affects the mechanical properties of the GM. The sieves were arranged in the following way according to their openings per linear inch: 4, 8, 16, 30, 50, 100, Pan.

The fineness modulus is given by:

$$\text{Fineness Modulus} = \frac{\Sigma \text{ Percent cumulative retained on sieve upto 160 micro meter}}{100} \quad (1)$$

The fineness modulus was calculated as 2.54, which was within limits. Fineness modulus for fine aggregate should be within the range 2.3 to 3.2.

3.2. Temperature of the Mortar

Temperature plays a crucial role in the hydration and curing process of construction materials. In this research, the temperature range of 15.6 °C to 26.7 °C suggests that the curing conditions were controlled within an acceptable range, promoting proper hydration and curing of the GM.

3.3. Setting Time of the Mortar

Initial setting times for different mortar mixes are presented in Table 2 below:

Table 2. Initial setting times of different mortars.

Sample ID	Initial Setting Time (min)
C10B90	40
C20B80	60
C30B70	75
OPC	35

3.4. Compressive Strength

Compressive strength for all the GM mixes were checked by testing them with the universal testing machine. A total of three cubes were tested for determination of strength for one day. The average of three cubes was taken as the strength of that day. Beside GM cubes, OPC mortar cubes were also tested for the sake of comparison. Compressive strengths of different mix designs are shown in Table 3. After compressive tests, it was found that the compressive strength of C20B80 cubes was the highest of all. It was 62 percent more than that of OPC at 28 days of curing. It can be seen in Table 3 below that the compressive strengths of all the GM cubes were greater than those of OPC mortar cubes.

Table 3. Compressive strengths of different mix designs.

GM Mix	Compressive Strength at Different Curing Ages (MPa)						
	7 Days	% Difference in Compressive Strength as Compared to OPC	14 Days	% Difference in Compressive Strength as Compared to OPC	28 Days	% Difference in Compressive Strength as Compared to OPC
C10B90	16.5	43.47	18.2	51.67	19.5	52.32
C20B80	18.1	57.39	19.6	63.34	20.76	62.18
C30B70	14.3	24.34	15.5	29.17	17.2	34.37
OPC	11.5	Nil	12.0	Nil	12.80	Nil

3.5. Cracks

It can be seen in Figure 1 that no cracks were observed in GM cubes. The absence of cracks in the GM cubes is a positive outcome, indicating that the mortar exhibited good structural integrity and resistance to cracking under the tested conditions. It suggests that the mixture proportions, including the binder-to-sand ratio and the incorporation of waste materials like corn cob ash and bauxite, were appropriate and resulted in a cohesive and robust mortar. Furthermore, the absence of cracks in the GM cubes may also be attributed to the curing conditions employed during the testing. The curing process, including the temperature and duration, likely played a significant role in allowing the mortar to develop its strength and preventing the formation of cracks.

Figure 1. Geopolymer cubes.

4. Conclusions

The following conclusions can be drawn from the conducted study:

1. One-part geopolymer can be developed locally in Pakistan by using corn cob ash and bauxite.
2. The mix design with 20 percent corn cob ash and 80 percent bauxite gave the most optimum results.
3. One-part geopolymer resulting from corn cob ash and bauxite has huge potential for replacing OPC.
4. The geopolymer developed in this research has compressive strength higher than that of normal ordinary Portland cement.

Author Contributions: Experimental methodology, research, R.A.; writing, editing and figures, A.I.; research, supervision, and writing—review, S.U.K. All authors have read and contributed to revising the paper. All authors have read and agreed to the published version of the manuscript.

Funding: This research received no external funding.

Institutional Review Board Statement: Not applicable.

Informed Consent Statement: Not applicable.

Data Availability Statement: Not applicable.

Acknowledgments: The authors express their gratitude to Muhammad Faisal Javed for his supervision and visualization of research.

Conflicts of Interest: The authors declare no conflict of interest.

References

1. Singh, R.P.; Vanapalli, K.R.; Cheela, V.R.S.; Peddireddy, S.R.; Sharma, H.B.; Mohanty, B. Fly Ash, GGBS, and silica fume based geopolymer concrete with recycled aggregates: Properties and environmental impacts. *Constr. Build. Mater.* **2023**, *378*, 131168. [CrossRef]
2. Lyddon, C. Focus on Pakistan. WorldGrain, 20-Sep-2021. Available online: https://www.world-grain.com/articles/15862-focus-on-pakistan (accessed on 7 May 2023).
3. Adesanya, D.A.; Raheem, A.A. Development of corn cob ash blended cement. *Constr. Build. Mater.* **2009**, *23*, 347–352. [CrossRef]
4. Ye, J.; Zhang, W.; Shi, D. Effect of elevated temperature on the properties of geopolymer synthesized from calcined ore-dressing tailing of bauxite and ground-granulated blast furnace slag. *Constr. Build. Mater.* **2014**, *69*, 41–48. [CrossRef]
5. *ASTM C136-05*; Standard Test Method for Sieve Analysis of Fine and Coarse Aggregates. ASTM: West Conshohocken, PA, USA, 2017. Available online: https://www.astm.org/c0136-05.html (accessed on 31 May 2023).
6. *ASTM C1064/C1064M*; Standard Test Method for Temperature of Freshly Mixed Hydraulic-Cement Concrete. ASTM: West Conshohocken, PA, USA, 2017. Available online: https://www.astm.org/c1064_c1064m-17.html (accessed on 31 May 2023).
7. *ASTM C191*; Standard Test Methods for Time of Setting Hydraulic Cement by Vicat Needle--eLearning Course. ASTM: West Conshohocken, PA, USA, 2010. Available online: https://www.astm.org/astm-tpt-185.html (accessed on 31 May 2023).
8. *ASTM C143/C143M*; Standard Test Method for Slump of Hydraulic-Cement Concrete. ASTM: West Conshohocken, PA, USA, 2015. Available online: https://www.astm.org/c0143_c0143m-12.html (accessed on 31 May 2023).
9. *ASTM C109/C109M*; Standard Test Method for Compressive Strength of Hydraulic Cement Mortars (Using 2-in. or [50-mm] Cube Specimens). ASTM: West Conshohocken, PA, USA, 2020. Available online: https://www.astm.org/c0109_c0109m-20.html (accessed on 31 May 2023).

Disclaimer/Publisher's Note: The statements, opinions and data contained in all publications are solely those of the individual author(s) and contributor(s) and not of MDPI and/or the editor(s). MDPI and/or the editor(s) disclaim responsibility for any injury to people or property resulting from any ideas, methods, instructions or products referred to in the content.

Proceeding Paper

Micro Structural Study of Concrete with Indigenous Volcanic Ash [†]

Muhammad Iqbal Bashir * and Ayub Elahi

Department of Civil Engineering, University of Engineering and Technology, Taxila 47080, Pakistan; ayub.elahi@uettaxila.edu.pk
* Correspondence: iqbal833@yahoo.com
† Presented at the 5th Conference on Sustainability in Civil Engineering (CSCE), Online, 3 August 2023.

Abstract: Extraordinary efforts should be carried out in Pakistan to prepare green concrete from waste materials. The utilization of Volcanic Ash (VA) in concrete can make sustainable concrete that will produce less carbon dioxide (CO_2) emissions and give positive outcomes. Hence, compressive strength was tested on VA concrete with changing concentrations ranging from 0, 10, and 20% with constant W/C, and the result was evaluated by scanning electron microscopy. The analysis of results reveals that the intrusion of VA with 10% replacement gives a significant response, and enhances the strength of the overall matrix.

Keywords: compressive strength; scanning electron microscope; volcanic ash; chemical composition

1. Introduction

Possible efforts should be carried out to use waste materials in concrete for the protection of the environment by making green concrete [1]. These waste materials may be agricultural, industrial, aquaculture, waste, natural minerals, dust powder, and ashes [2]. Volcanic Ash concrete can be considered green concrete. Moreover, SCMs utilization can also improve mechanical and durability properties. The commonly used SCM includes Volcanic Ash (VA). By adding SCMs, overall mechanical and durability properties are improved. Their utilization in concrete consumes less energy for production and evolves less CO_2. Moreover, protection against freeze and thaw, alkali–silica reaction, chloride attack, and sulfate attack may also be achieved. Siddique et al. [3] revealed that the compressive strength was diminished as the proportion of Volcanic Ash (VA) replacement in cement increased. A decrease of around 40% in strength was observed when 40% of the cement was substituted with VA. Mostafa et al. [4] investigated the effect of the VA with and without magnetizing water (MW). The author used different concentrations of VA with cement replacement and revealed that the VA with 15% depicts a significant increase of about 33% in strength. Anwar et al. [5] investigated the impact of the partial replacement of cement with Volcanic Ash (VA) and pumice powder (VP) on the compressive strength of cement mortar. The replacement percentage ranged from 0 to 50%, and tests were carried out over 28 days. Based on the findings, it was observed that the compressive strength decreased as the content of VA or VP increased. This decrease in strength can be attributed to the reduction in the amount of cement in the mixture due to the higher content of VA or VP. Ekinci et al. [6] discovered that the addition of volcanic material to geopolymer concrete resulted in decreased workability, which in turn can negatively affect the compressive strength of the concrete. Moreover, a scientometric diagram, as shown in Figure 1, depicts the importance of Volcanic Ash in concrete.

There is some research on sustainable concrete using VA. However, there is minimum data related to microstructural studies available to verify exhibited mechanical properties. This research aims to use locally available Volcanic Ash as a partial replacement for cement to make sustainable concrete without compromising on compressive, and an attempt will be made to verify these results through microstructural studies of concrete with VA.

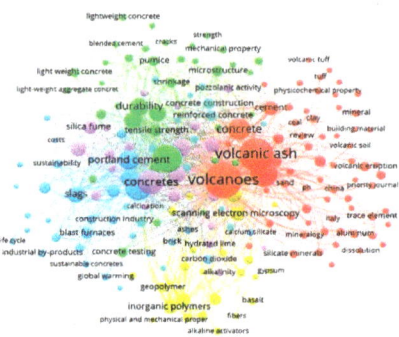

Figure 1. Scientometric diagram of Volcanic Ash.

2. Experimental Procedures

2.1. Material Used

2.1.1. Volcanic Ash

VA as a waste material was taken from the locally available place near Chilas (Pakistan), and its chemicals analysis reveals that VA has similar contents and a greater concentration of SiO_2 than OPC, as illustrated in Table 1. The $SiO_2 + Al_2O_3 + Fe_2O_3$ concentration is more than 70%. This depicts the pozzolanic nature of VA as per ASTM C618-01. In addition, Table 1 also demonstrates the physical composition of VA.

Table 1. The chemical composition of VA.

Chemical Composition		Physical Composition	
Oxide	VA (%Age by Mass)	Characteristics	VA
SiO_2	53.69	Specific Gravity	2.67%
Al_2O_3	17.43	Soundness	No Expansion
Fe_2O_3	9.52	Retain on sieve # 325 max (%)	33
CaO	7.00		
MgO	3.87		
Na_2O	3.57		
K_2O	0.86		
SO_3	0.16		
Lime saturation Factor	3.89		
Silica Modulus	1.99		
Aluminum Modulus	1.83		
L.O. I	1.3		

2.1.2. Cement

Ordinary Portland Cement (OPC) was used and had a chemical and physical composition as per ASTM C-150 type-1 (normal).

2.1.3. Coarse/Fine Aggregate

Margalla crush, with a max size equal to 3/4″, and Qibla Bandy sand were used for making the concrete mix.

2.2. Concrete Mix Proportion

Concrete mix proportions are shown in Table 2.

Table 2. Concrete mix proportions.

Mixes	Concrete Mix Composition				
	Cement (Kg/m^3)	Volcanic Ash (Kg/m^3)	Water (w/c = 0.5) (Kg/m^3)	Fine Aggregate (Kg/m^3)	Coarse Aggregate (Kg/m^3)
Control Sample	320	0	160	640	1280
V10	288	32	160	640	1280
V20	256	64	160	640	1280

Sample Preparation

Mixing of concrete was performed with w/c = 0.5. Homogeneously mixed samples were cast in 4″ Ø, 8″ long cylinders in three layers of compaction. The cast samples were de-molded after 24 h and then cured with a wet hessian cloth that was maintained at room temperature of 25 °C + 3 °C. Mix proportions are included in Table 2.

2.3. Tests Performed

Compressive Strength and Scanning Electron Microscopy of the Concrete

Compressive strength was carried out after 28× days. The average value of the three specimens for each test was determined and recorded. Compressive strength was evaluated on the bases of ASTM C039, and SEM was performed to check the internal microscopy of the structure.

3. Results and Discussions

3.1. Compressive Strength and Microstructure of VA Concrete 1:2:4

Compressive strength values of various mixes with varying concentrations are shown in Figure 2. The best result of compressive strength was achieved for the mix VA-10 compared to that of normal concrete. Moreover, there was a decrease in strength observed with a higher concentration of VA, as demonstrated in Figure 2a. Initially, the increase in strength was due to the pozzolanic hydration process between cement and VA. The pozzolanic reaction between Volcanic Ash and CH produced additional Calcium Silicate Hydrate (C-S-H) and produced dense gel. Volcanic Ash also reacted with CH and aluminates to form C-A-S-H gel. This provided additional strength as C-A-S-H is denser than CH. Thus, it contributed to the densification of the concrete structure. The SEM analysis reveals that the volcanic ash particles, when mixed with cement, form a highly compact and dense C-S-H gel. This is the primary binding material in concrete, responsible for its strength and durability. The volcanic ash particles interact with the cement, promoting the formation of additional C-S-H gel. The denser gel structure contributes to the overall strength of the concrete, as illustrated in Figure 2b.

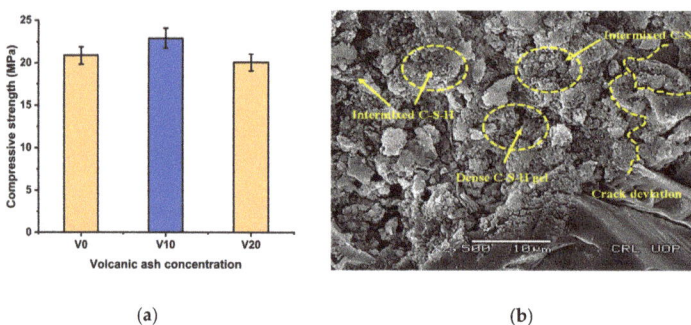

Figure 2. (**a**) Compressive strength of VA; (**b**) scanning electron microscopy of VA with 10%.

3.2. Cost Benefit Analysis

A system process to evaluate suitability by weighing its potential benefits and cost is called cost-benefit analysis. The rate of normal PCC 1:2:4 in the foundation without shuttering for the 1 m^3 has been compared, as illustrated in Table 3. This comparison is specifically for the province of Gilgit Baltistan:

Table 3. Cost comparison of VA (20%) replacement with P.C.C.

	P.C.C Control Sample			P.C.C with 10% Replacement		
Parameters	Quantity	Rates	Amount	Quantity	Rates	Amount
Cement	6.4 Bags	Rs. 1350/Bag	Rs. 8640/-	5.12 Bags	Rs. 1350/Bag	Rs. 6912/-
VA	-	-	-	64 Kg	Rs. 3/Kg	Rs. 192/-
Sand	16 ft^3	Rs. 80/ft^3	Rs. 1280/-	16 ft^3	Rs. 80/ft^3	Rs. 1280/-
Crush	32 ft^3	Rs. 110/ft^3	Rs. 3520/-	32 ft^3	Rs. 110/ft^3	Rs. 3520/-
Labor for pouring and curing	35.311 ft^3	Rs. 30/ft^3	Rs. 1059/-	35.311 ft^3	Rs. 30/ft^3	Rs. 1059/-
Total			14,499/-			12,963/-
Cost reduction			Rs. 1536/Cum (11%)			

4. Conclusions

A comprehensive study has been carried out by replacement of cement with concrete. The following are the conclusions from this comprehensive study:

1. The compressive strength of concrete with 10% VA replacement enhances the composite strength compared to the control specimen;
2. SEM analysis reveals that VA particles react with the CH to form densified C-S-H gel. In addition, deviation of cracks is observed, which is a good sign for strength and durability;
3. Incorporating Volcanic Ash (VA) in concrete construction leads to a significant cost reduction of 11% when considering the desired compressive strength.

Author Contributions: M.I.B.: writing, investigation, methodology, and drafting; A.E.: supervision, resources. All authors have read and agreed to the published version of the manuscript.

Funding: This research received no external funding.

Institutional Review Board Statement: Not applicable.

Informed Consent Statement: Not applicable.

Data Availability Statement: Not applicable.

Acknowledgments: I would like to express my sincere appreciation to my colleagues Shaheer Ahmad Janjua, Furqan Farooq, and Samaha Badi Uz-Zaman, and classmates for their intellectual discussions, valuable insights, and continuous support.

Conflicts of Interest: The authors declare no conflict of interest.

References

1. Akbar, A.; Farooq, F.; Shafique, M.; Aslam, F.; Alyousef, R.; Alabduljabbar, H. Sugarcane bagasse ash-based engineered geopolymer mortar incorporating propylene fibers. *J. Build. Eng.* **2021**, *33*, 101492. [CrossRef]
2. AlKhatib, A.; Maslehuddin, M.; Al-Dulaijan, S.U. Development of high performance concrete using industrial waste materials and nano-silica. *J. Mater. Res. Technol.* **2020**, *9*, 6696–6711. [CrossRef]
3. Siddique, R. Properties of concrete made with volcanic ash. *Resour. Conserv. Recycl.* **2012**, *66*, 40–44. [CrossRef]
4. Keshta, M.M.; Yousry Elshikh, M.M.; Kaloop, M.R.; Hu, J.W.; ELMohsen, I.A. Effect of magnetized water on characteristics of sustainable concrete using volcanic ash. *Constr. Build. Mater.* **2022**, *361*, 129640. [CrossRef]

5. Anwar Hossain, K.M. High strength blended cement concrete incorporating volcanic ash: Performance at high temperatures. *Cem. Concr. Compos.* **2006**, *28*, 535–545. [CrossRef]
6. Alqarni, A.S. A comprehensive review on properties of sustainable concrete using volcanic pumice powder ash as a supplementary cementitious material. *Constr. Build. Mater.* **2022**, *323*, 126533. [CrossRef]

Disclaimer/Publisher's Note: The statements, opinions and data contained in all publications are solely those of the individual author(s) and contributor(s) and not of MDPI and/or the editor(s). MDPI and/or the editor(s) disclaim responsibility for any injury to people or property resulting from any ideas, methods, instructions or products referred to in the content.

Proceeding Paper

Flexure Response of Stainless-Steel-Reinforced Concrete (SSRC) Beams Subjected to Fire [†]

Javaria Mehwish [1,*], Katherine A. Cashell [2] and Rabee Shamass [3]

1. Department of Civil and Environmental Engineering, Brunel University London, London UB8 3PH, UK
2. Department of Civil Environmental and Geomatic Engineering, University College London (UCL), London WC1E 6BT, UK; k.cashell@ucl.ac.uk
3. Division of Civil and Building Services Engineering, School of Built Environment and Architecture, London South Bank University (LSBU), 103 Borough Road, London SE1 0AA, UK; shamassr@lsbu.ac.uk
* Correspondence: javaria.mehwish@brunel.ac.uk
† Presented at the 5th Conference on Sustainability in Civil Engineering (CSCE), Online, 3 August 2023.

Abstract: This paper examines the behavior of stainless-steel-reinforced concrete (SSRC) flexural members subjected to fire. Stainless steel (SS) reinforcement has gained popularity due to its corrosion resistance and long maintenance-free life. However, there is an insufficiency of performance data and design guidance in the present literature. This paper presents a numerical assessment of SSRC structural elements using a material model based on experimental tests. A finite element model was utilized to simulate and analyze the response of SSRC beams under fire. This study compared the behavior of SSRC beams with traditional carbon-steel-reinforced concrete (CSRC) beams, demonstrating that SSRC members have a higher load carrying capacity and can sustain fire exposure for longer durations. Additionally, SSRC beams exhibited higher deflections during fire exposure compared to CSRC beams.

Keywords: ABAQUS; finite element modeling; stainless steel; reinforced concrete

1. Introduction

In past few years, there has been a noticeable surge in the utilization of stainless steel (SS) rebars as an appealing substitute for traditional carbon steel reinforcement in the United Kingdom. This growing trend can be attributed to its advantageous and sustainable characteristics, including exceptional resistance to corrosion, leading to extended periods of maintenance-free durability. However, there remains a significant gap in publicly available data concerning the performance and design aspects, particularly in extreme circumstances like fires. While SS rebars were primarily employed in bridges and structures such as water treatment plants, which were susceptible to corrosion, their application has broadened to encompass various structural uses in recent years, including industrial buildings, car parks, and marine environments.

Despite the extensive research on various aspects of SS rebars, there is a notable scarcity of information regarding its behavior at elevated temperatures. Figure 1a,b depict the retention factors for yield strength and Young's modulus at different elevated temperature levels for bare SS structural sections, comparing the performance to both carbon steel (CS) [1] and grade 1.4301 [2,3] stainless steel. These distinct properties of SS prove highly beneficial during fire incidents. However, it is important to note that stainless steel's higher coefficient of linear thermal expansion (between $14–17 \times 10^{-6}/°C$) compared to carbon steel ($12 \times 10^{-6}/°C$) presents a challenge in maintaining the bond between SS rebars and the surrounding concrete under elevated temperature scenarios. This can lead to compromised composite action, increased cracking, and heightened levels of concrete spalling.

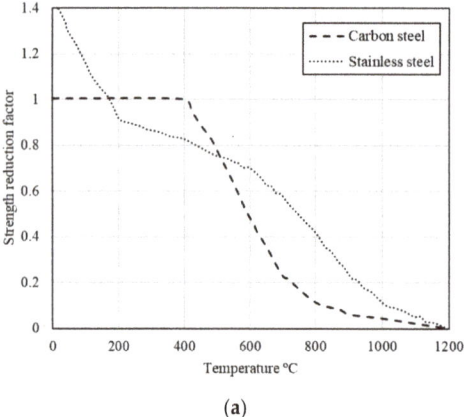
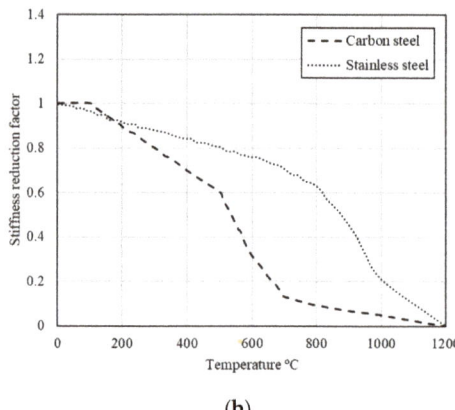

(a)   (b)

Figure 1. Retention of mechanical properties for stainless steel (SS) and carbon steel (CS) including (**a**) strength; (**b**) stiffness (adapted from Ref. [3]).

In light of this dearth of data, the current study aimed to address the elevated temperature behavior of SSRC in fire. The approach involves the development and validation of a finite element model, incorporating experimental data [4], to accurately represent the material properties of SS rebars at elevated temperatures.

2. Finite Element Analysis

2.1. General

A numerical model was developed to simulate and study the behavior of SSRC structural members at elevated temperatures. To date, there is no physical test data for this type of structural behavior. In terms of the ambient temperature behavior, a number of researchers have conducted a numerical analysis of SSRC beams (e.g., [5–7]). Therefore, the numerical model developed in this paper was validated using CSRC beams tested by Dwaikat and Kodur [8].

2.2. Structural Arrangement

The numerical model was developed based on a sample beam with the details of Beam B-1, which was examined by Dwaikat and Kodur [8] under standard fire curve ASTM E119 [9], and was made of normal strength concrete with a 58 MPa compressive strength. As shown in Figure 2, the simply supported beam was 3960 mm in length, 254 mm in width, and had a total depth of 406 mm. The beam had tensile reinforcement from three 19 mm bars and compression reinforcement from two 13 mm bars. Shear reinforcement was also included in the cross-section and this was 6 mm bars at a constant spacing of 150 mm. The nominal yield strength of the longitudinal rebar was 420 N/mm^2 and 280 N/mm^2 for the stirrups. The beam was loaded in 4-point loading conditions; the two loading points were 1200 mm apart from each other.

2.3. Material Modeling

The concrete material behavior is defined through the damage plasticity material model (CDP), available in the ABAQUS [10] library. The CDP model includes the effect of elevated temperature and can model the inelastic response of concrete in both tension and compression. The concrete material model was developed using Eurocode 2 [2].

An isotropic yielding of steel reinforcement with temperature dependence can be defined in ABAQUS [10] by a uniaxial yield surface against uniaxial plastic strain. A constitutive material model of the carbon steel reinforcement was also taken from Eurocode 2 [2]. In the current study, actual stress–strain curves for both austenitic and duplex steel, determined through practical tests [4] and shown in Figures 2 and 3, were implemented in the ABAQUS model for validation at elevated temperatures.

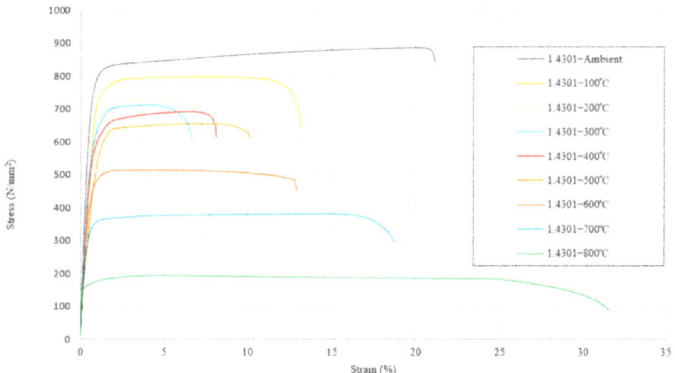

Figure 2. Stress-strain responses for grade 1.4301 at elevated temperatures adapted from Ref. [4].

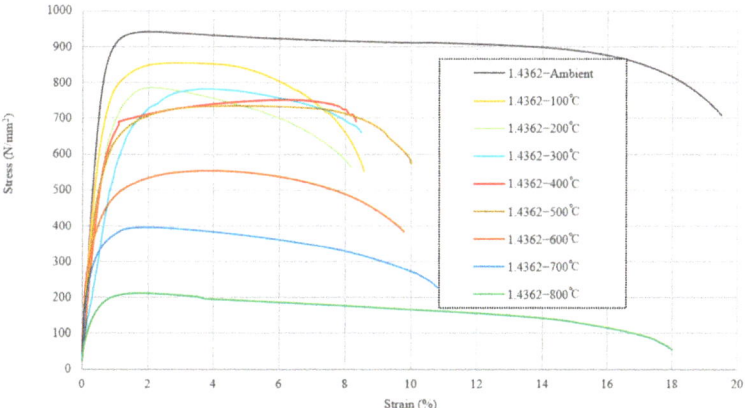

Figure 3. Stress–strain responses for grade 1.4362 at elevated temperature adapted from Ref. [4].

2.4. Sequentially Coupled Thermal Stress Analysis

There are generally two different approaches in finite element analysis for the solution of structural fire analyses. There are fully coupled and sequentially coupled thermal-stress analyses. The former of these is hugely computationally demanding and therefore, for RC structures, a sequentially coupled stress analysis is typically employed as it is more computationally efficient; this was employed in the current work. A sequentially coupled thermal-stress analysis is performed in two steps: (i) a heat transfer analysis is first conducted to simulate the spread of an elevated temperature through the sections and (ii) a thermal-stress analysis is then performed to apply the thermal loads and compute displacements. The temperature at each node of the element is calculated in the first step and these are then applied as a predefined field in the second step. Details of these steps are given in another paper by the author that was published earlier [11].

2.5. Validation of the Numerical Model

The validated numerical model was employed for the simulation of SSRC beams under fire. The only difference in the model is the material model for the reinforcement. The interaction between the rebar and the concrete is modeled as a perfect bond in both cases; however, the tensile stiffening modeled for concrete allows for modeling the effect of interaction in a simple manner [10]. In the current study, stress–strain curves measured in practical tests for both austenitic (1.4301) steel and duplex (1.4362) steel at elevated temperatures were used in the numerical model to determine the fire resistance of SSRC beams under fire. There was a wide difference in the thermal expansion of CS and SS rebars, but it did not result in any changes in the heat distribution between the concrete and rebars (Figure 4).

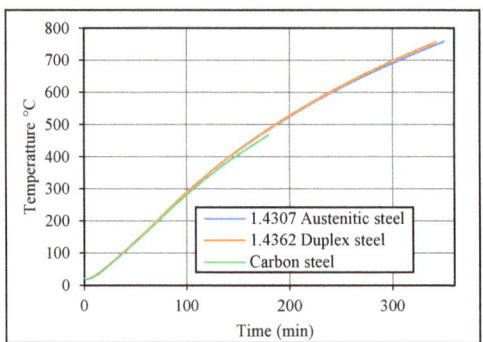

Figure 4. Comparison of temperature in reinforcement of similar RC beams reinforced with SS and CS.

2.6. Results

The comparison of two RC beams reinforced with CS and austenitic steel with a similar cross-sectional geometry, concrete strength, and reinforcement ratio was performed for sample beam validation, is shown in Figure 5. The SSRC beam showed no failure under the ASTM E119 [9] fire curve, allowing for a reduced reinforcement ratio to analyze its behavior under the full-time–temperature curve. A further comparison showed that SSRC sustains loads under fire much longer than CS, even with a reduced reinforcement ratio. The CSRC beam failed at around 175 min, while the austenitic and duplex SSRC beams failed at 237 and 243 min, respectively, with different failure modes: concrete crushing for SSRC beams and a combination of concrete crushing and steel rupture for CSRC beams.

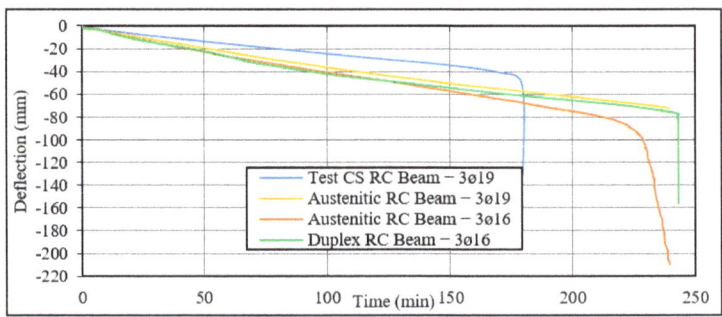

Figure 5. Fire resistance of RC beams reinforced with different steel ratios of CS, austenitic, and duplex steel.

3. Conclusions

This study aimed to examine the mechanical behavior of austenitic and duplex steel rebars when exposed to elevated temperatures. To achieve this, actual material models, developed through a testing program, were incorporated into a validated finite element model of an SSRC beam that was subjected to a fire scenario.

The numerical results obtained from the analysis indicated that the SSRC beams were able to withstand fire for a much longer duration compared to CSRC beams. This suggests that stainless steel exhibits superior fire resistance properties. Furthermore, the SSRC beams experienced higher deflections during the fire exposure when compared to CSRC beams.

Based on these findings, it is recommended that further research studies could be conducted to investigate the behavior of SSRC beams at elevated temperatures with different material strength and load ratios. This would provide a more comprehensive understanding of the performance of SSRC structures in fire scenarios, allowing for the development of more accurate design guidelines and safety standards.

Author Contributions: Conceptualization, K.A.C.; Data curation, J.M.; Formal analysis, J.M., K.A.C. and R.S.; Funding acquisition, J.M.; Investigation, J.M.; Methodology, J.M. and K.A.C.; Project administration, K.A.C. and R.S.; Resources, J.M., K.A.C. and R.S.; Software, J.M. and R.S.; Supervision, K.A.C. and R.S.; Validation, J.M., K.A.C. and R.S.; Visualization, J.M.; Writing—original draft, J.M.; Writing—review & editing, J.M., K.A.C. and R.S. All authors have read and agreed to the published version of the manuscript.

Funding: This research was funded by the UKRI, under an EPSRC Standard Research Studentship (DTP). The funding training grant number is EP/R512990/1.

Institutional Review Board Statement: Not applicable.

Informed Consent Statement: Not applicable.

Data Availability Statement: The data that support the findings of this study are available from the corresponding author, [J.M. and K.A.C.], upon reasonable request.

Conflicts of Interest: The authors declare no conflict of interest.

References

1. *BS EN 1993-1-1*; Eurocode 3: Design of Steel Structures—Part 1.1: General Rules and Rules for Buildings. BSI Group: London, UK, 2006.
2. *BS EN 1992-1-2*; Eurocode 2: Design of Concrete Structures–Part 1-2: General Rules-Structural Fire Design. BSI Group: London, UK, 2004.
3. *BS EN 1993-1-4, 2006 +A1*; Eurocode 3: Design of Steel Structures—Part 1.4: General Rules—Supplementary Rules for Stainless Steels. BSI Group: London, UK, 2015.
4. Rehman, F.; Cashell, K.A.; Anguilano, L. Experimental Study of the Post-Fire Mechanical and Material Response of Cold-Worked Austenitic Stainless Steel Reinforcing Bar. *Materials* **2022**, *15*, 1564. [CrossRef] [PubMed]
5. Alih, S.; Khelil, A. Behavior of inoxydable steel and their performance as reinforcement bars in concrete beam: Experimental and nonlinear finite element analysis. *Constr. Build. Mater.* **2012**, *37*, 481–492. [CrossRef]
6. Franco, N.; Biscaia, H.; Chastre, C. Experimental and numerical analyses of flexurally-strengthened concrete T-beams with stainless steel. *Eng. Struct.* **2018**, *172*, 981–996. [CrossRef]
7. Musab Rabi, K.A. Cashell, Rabee Shamass, Ultimate behaviour and serviceability analysis of SSRC. *Eng. Struct.* **2021**, *248*, 113259. [CrossRef]
8. Dwaikat, M.B.; Kodur, V.K.R. Response of restrained concrete beams under design fire exposure. *J. Struct. Eng.* **2009**, *135*, 1408–1417. [CrossRef]
9. *ASTM E119*; Standard Methods of Fire Test of Building Construction and Materials. American Society of Testing and Materials: West Conshohocken, PA, USA, 2008.
10. *ABAQUS*, 2020; Finite Element Software; User Assistance Manual; Dassault Systèmes; Simulia: Providence, RI, USA, 2020.
11. Mehwish, J.; Cashell, K.A. Rabee Shamass, 3D finite element analysis of reinforced concrete beams under fire. In Proceedings of the 4th Conference on Sustainability in Civil Engineering (CSCE'22), Islamabad, Pakistan, 31 August 2022.

Disclaimer/Publisher's Note: The statements, opinions and data contained in all publications are solely those of the individual author(s) and contributor(s) and not of MDPI and/or the editor(s). MDPI and/or the editor(s) disclaim responsibility for any injury to people or property resulting from any ideas, methods, instructions or products referred to in the content.

MDPI
St. Alban-Anlage 66
4052 Basel
Switzerland
www.mdpi.com

Engineering Proceedings Editorial Office
E-mail: engproc@mdpi.com
www.mdpi.com/journal/engproc

Disclaimer/Publisher's Note: The statements, opinions and data contained in all publications are solely those of the individual author(s) and contributor(s) and not of MDPI and/or the editor(s). MDPI and/or the editor(s) disclaim responsibility for any injury to people or property resulting from any ideas, methods, instructions or products referred to in the content.

www.ingramcontent.com/pod-product-compliance
Lightning Source LLC
LaVergne TN
LVHW070044120526
838202LV00101B/424